Virtual Equivalent System Approach for Stability Analysis of Model-based Control Systems

Weicun Zhang · Qing Li

Virtual Equivalent System Approach for Stability Analysis of Model-based Control Systems

Springer

Weicun Zhang
School of Automation
and Electrical Engineering
University of Science
and Technology Beijing
Beijing, China

Qing Li
School of Automation
and Electrical Engineering
University of Science
and Technology Beijing
Beijing, China

ISBN 978-981-15-5540-4 ISBN 978-981-15-5538-1 (eBook)
https://doi.org/10.1007/978-981-15-5538-1

This Springer imprint is published by the registered company Springer Nature Singapore Pte Ltd.
The registered company address is: 152 Beach Road, #21-01/04 Gateway East, Singapore 189721, Singapore

To our families and friends for their love and support!

Preface

Based on theoretical analysis and simulation results, a new concept of Virtual Equivalent System (VES) is proposed in recent years. The new concept enables us to develop a unified framework for analyzing the stability and convergence of Self-Tuning Control (STC) systems, or even more, of all adaptive control systems, without considering more details about the related estimation algorithms and control schemes.

With the aid of the VES concept, we can convert a time-varying STC system into a time-invariant system with a certain nonlinear compensation signal, whereby to reduce the complexity and difficulty in stability and convergence analysis. By doing so, we actually convert the original nonlinear dominant (nonlinear in structure) problem to a new equivalent problem which is linear dominant one (linear in structure). In return for such disposal, some more concise and applicable propositions are obtained.

After that, the VES concept and methodology are used to analyze the stability of Multiple Model Adaptive Control (MMAC) systems and T-S model based fuzzy control systems. In future research, we are planning to analyze the stability of other model-based control systems, such as Model Predictive Control (MPC).

Beijing, China
August 2019

Weicun Zhang
Qing Li

Acknowledgments

This monograph was supported in part by the National Natural Science Foundation of China (No. 61520106010, No. 61741302)

Contents

Chapter 1
Motivation and Contents of the Monograph

1.1 Brief History of Adaptive Control

From the viewpoint of application, adaptive control was originated from gain-scheduling control of high-performance aircraft in the early 1950s. Model reference adaptive control (MRAC) was suggested by Whitaker et al. [1] to solve the autopilot control problem. From the viewpoint of theory research of optimal control of stochastic system with unknown or time varying parameters, self-tuning control (STC) was suggested by Kalman [2], and then connected with applications through the pioneering work of Astrom and Wittenmark [3]. It has been well-known that the model reference control is actually equivalent to STC. To be specific, MRAC can be considered as a special class of STC. The only difference is that MRAC was first developed for continuous-time plants for model following, whereas STC was initially developed for discrete-time plants in a stochastic environment using minimization techniques [4–9].

The lack of stability analysis and the lack of understanding of the properties of various adaptive control schemes coupled with a disaster in a flight test caused the interest in adaptive control to diminish in 1960s. Especially, the Rohr's counter-example to robustness of adaptive control put forwarded severe challenges to the theory of adaptive control in early 1980s.

To address the above-mentioned theoretical problems, many corresponding research results have been published on stability, convergence, and robustness of adaptive control schemes, as well as many engineering applications.

Despite the fundamental progress achieved so far, a general theory of STC is still absent. As a rule, a well-developed theory should be able to provide deep insights and understanding on the mechanism and properties of a STC system, as well as useful guidelines to the design of new self-tuning schemes for various systems encountered in applications. But, just as Artrom and Wittenmark pointed out in their book: Adaptive Control, "unfortunately, there is no collection of results that can be called a

W. Zhang and Q. Li, *Virtual Equivalent System Approach for Stability Analysis of Model-based Control Systems*, https://doi.org/10.1007/978-981-15-5538-1_1

theory of adaptive control in the sense specified." In the literature, many of the existing works have been specific to particular estimation algorithms and control schemes, while only a few research efforts have been made towards a unified analysis of the subject from a general perspective [10–13].

This monograph was motivated by these situations to present a unified analysis tool, i.e., Virtual Equivalent System (VES), to understand and judge the stability, convergence, and robustness of STC, and other model based control schemes.

1.2 Contents of the Monograph

This monograph includes 7 chapters and 4 appendixes.

Chapter 1: "Motivation and Contents of the Monograph" gives a brief history of adaptive control, motivation and contents of the monograph, and some new perspectives in this monograph;

Chapter 2: "Stability and Convergence Analysis of Self-Tuning Control Systems" introduces virtual equivalent system(VES) concept and methodology with the analysis of STC systems under assumption that structure information of the plant to be controlled is known;

Chapter 3: "Further Results On Stability and Convergence of Self-Tuning Control" gives further corresponding results without requiring the structure information of the plant to be controlled;

Chapter 4: "Stable Weighted Multiple Model Adaptive Control of Discrete-Time Stochastic Plant" adopts VES concept and methodology to analyze the closed-loop stability of weighted multiple model adaptive control(WMMAC) of discrete-time stochastic plant;

Chapter 5: "Further Results on Stable Weighted Multiple Model Adaptive Control of Discrete-Time Stochastic Plant" gives further corresponding results with improved weighting algorithm;

Chapter 6: "Stable Weighted Multiple Model Adaptive Control of Continuous-Time Plant" adopts VES concept and methodology to analyze the closed-loop stability of weighted multiple model adaptive control(WMMAC) of continuous-time plant; It is not trivial to prove the closed-loop stability of continuous-time WMMAC system, although the stability of discrete-time WMMAC system has been proved in Chaps. 4 and 5 for different situations.

Chapter 7: "Stability of Continuous-Time T-S Model Based Fuzzy Control Systems" extends the VES concept and methodology to stability analysis of T-S model based fuzzy control systems.

1.3 Some New Perspectives

With the help of VES concept and methodology, this monograph delivers the following new perspectives:

1. Persistent excitation is not necessary for self-tuning adaptive control objective, except for requiring the convergence of parameter estimation. In other words, feedback information of adaptive control is sufficient to achieve the control objective (regulation or tracking).
2. STC system based on suitable low order model may also achieve the desired control objective, i.e., stable and convergent, if only the plant (system) to be controlled can be approximated by suitable low order model, say second-order model for most engineering applications.
3. A unified or general analysis theory and corresponding criteria are presented for different kinds of STC schemes. To some extent, VES can give an explicit explanation why an STC system could be stable and convergent without requiring the convergence or consistent convergence of parameter estimate.

In particular, with the help of VES concept and corresponding stability and convergence criteria, we wanted to explain the "surprising" of Astrom and Wittenmark, which was expressed in the abstract of their famous paper [3], "It is shown that if the parameter estimates converge the control law obtained is in fact the minimum variance control law that could be computed if the parameters of the system were known. This is somewhat surprising since the least squares estimate is biased."

References

1. H.P. Whitaker, J. Yamron, A. Kezer, *Design of Model Reference Adaptive Control Systems for Aircraft, Report R-164, Instrumenta- tion Laboratory* (M. I. T. Press, Cambridge, Massachusetts, 1958)
2. R.E. Kalman, Design of a self optimizing control system. Trans. ASME **80**, 468–478 (1958)
3. K.J. Åström, B. Wittenmark, On self-tuning regulators. Automatica **9**, 185–199 (1973)
4. S. Sastry, M. Bodson, *Adaptive Control: Stability, Convergence and Robustness* (Prentice Hall, Englewood Cliffs, New Jersey, 1989)
5. P.A. Ioannou, J. Sun, *Robust Adaptive Control* (Prentice-Hall: Englewood Cliffs, New Jersey, USA, 1996)
6. B. Egardt, Unification of some discrete time adaptive control schemes. IEEE Trans. Automatic Control **25**, 693–697 (1980)
7. P.J. Gawthrop, Some interpretations of the self-tuning controller. Proc. IEEE **124**, 889–894 (1977)
8. L. Ljung, I.D. Landau, Model reference adaptive systems and self-tuning regulators—some connections, in *Proceedings of 7th IFAC Congress*, vol. 3 (1978), pp. 1973–1980
9. K.S. Narendra, L.S. Valavani, Direct and indirect adaptive control. Automatica **15**, 663–664 (1979)
10. J.H. Van Schuppen, Tuning of gaussian stochastic control systems. IEEE Trans. Automatic Control **39**(11), 2178–2190 (1994)

11. K. Nassiri-Toussi, W. Ren, A unified analysis of stochastic adaptive control: asymptotic self-tuning, in *Proceedings of the 34th IEEE Conference on Decision and Control* (1995), pp. 2932–2937
12. A.S. Morse, Towards a unified theory of parameter adaptive control-part II: certainty equivalence and implicit tuning. IEEE Trans. Automatic Control **37**(1), 15–29 (1992)
13. K. Nassiri-Toussi, W. Ren, Indirect adaptive pole-placement control of MIMO stochastic systems: self-tuning results. IEEE Trans. Automatic Control **42**(1), 38–52 (1997)

Chapter 2
Stability and Convergence Analysis of Self-Tuning Control Systems

Abstract This chapter presents the concept of virtual equivalent system (VES) and corresponding criteria of the stability and convergence of a general STC system consisted of arbitrary control strategy and arbitrary estimation algorithm. The necessary conditions required for global stability and convergence of a general STC system (not only minimum variance STC system for minimum-phase plant) are relaxed, i.e. the convergence of parameter estimates is removed. The plants to be controlled include deterministic and stochastic, linear time-invariant (LTI) and linear time-varying (LTV). With the help of VES, the original nonlinear dominant (nonlinear in structure) problem is converted to a linear dominant problem (linear in structure). The main results are expressed by 10 theorems plus some new perspectives on STC.

2.1 Introduction

After more than half-century research, a lot of achievements of STC systems have been made. Many research efforts have been devoted to develop stable and convergent STC schemes, see, e.g., Refs. [1–12], and the references therein. For more details, the reader is referred to some well known books, such as [13–18]. As representative works in the proof of stability and convergence of STC systems, Ref. [5] proved that the STC system comprising stochastic gradient (SG) estimates and minimum-variance control is stable and convergent without the convergence of parameter estimates; Ref. [10] obtained a similar conclusion for the STC system based on recursive extended least-square (ELS) estimates and minimum-variance control.

Although fundamental progress have been made, a general theory of STC is still absent. There is still a long way to go from having a full understanding of this important class of control strategies [19]. Some similar remarks could also be found in [17, 20]. In the bulk of literature, almost all the existing works have been specific to particular estimation algorithms and control schemes, while only a few attempts have been made towards a unified analysis of the subject from a general perspective, e.g., [9, 21–23].

To strengthen the above-mentioned viewpoints, here we put forward a question that cannot be answered by the existing theory. Why can minimum variance (MV)

W. Zhang and Q. Li, *Virtual Equivalent System Approach for Stability Analysis of Model-based Control Systems*, https://doi.org/10.1007/978-981-15-5538-1_2

self-tuning control attain the stability and convergence without the convergence of parameter estimates, but other self-tuning schemes, like pole placement or LQG [24, 25], cannot? It is clear that additional theoretical research is needed to expand the classes for which stability can be proven; thus, relaxation of the necessary conditions and extension to even more general classes are still to be investigated [26].

In the authors opinion, such situation is mainly due to the lack of appropriate methodology and corresponding analysis tool. Motivated by this observation, the VES concept and corresponding analysis approach is proposed. The idea of VES was originated from Ph.D thesis of Weicun Zhang [27], and then gradually improved later in References [28–32]. VES is actually an artificial system equivalent to the STC system in the input-output sense. From another point of view, through VES we convert the original nonlinear dominant (nonlinear in structure) system to a linear dominant system (linear in structure). In return for such disposal, as we will see later, the analysis of stability and convergence of STC systems becomes more direct and easy-understandable. Furthermore, the consistency or the convergence of parameter estimates is removed from the necessary conditions for the stability and convergence of a general STC system.

Additionally, it has been well-known that the model reference adaptive control, another important ingredient of adaptive control, is actually equivalent to STC [15, 17, 33–36]. Therefore, a unified/general theory of stability and convergence of adaptive control can be expected.

To avoid ambiguity with the convergence of parameter estimation, it is necessary to emphasize here that the "convergence" of an STC system means its performance index converges to the corresponding non-adaptive control system.

2.2 Deterministic Self-Tuning Control System

2.2.1 *Construction of Virtual Equivalent System*

Three kinds of VES will be constructed for analyzing deterministic STC systems. First, we present two of them according to two different situations of parameter estimates: convergent to true values, convergent but not consistent. These two types of VES are comparatively easy to understand. As for the situation that the parameter estimates may not converge, we will use another type of VES, which is a slow switching system. It will be described later in the proof of Theorem 2.4 in Sect. 2.2.2.3.

Consider the following deterministic time-invariant plant **P**

$$A(q^{-1})y(k) = q^{-d}B(q^{-1})u(k) \tag{2.1}$$

with

$$A(q^{-1}) = 1 + a_1 q^{-1} + \cdots + a_n q^{-n}$$
$$B(q^{-1}) = b_0 + b_1 q^{-1} + \cdots + b_m q^{-m}$$

where $y(k)$ and $u(k)$ are the system output and input respectively, $y(k) = 0$, $u(k) = 0$ for $k < 0$, $A(q^{-1})$ and $B(q^{-1})$ are polynomials in backward-shift operator q^{-1} with known orders or upper bounds of orders n, m, and unknown coefficients.

Equation (2.1) can be rewritten as

$$y(k) = \phi^T(k-d)\theta \tag{2.2}$$

where

$$\phi^T(k-d) = [y(k-1), \ldots, y(k-n), u(k-d), \ldots, u(k-d-m)] \tag{2.3}$$

$$\theta^T = [-a_1, \ldots - a_n, b_0, b_1, \ldots, b_m] \tag{2.4}$$

The estimates of θ are denoted by $\hat{\theta}(k)$, which corresponds to model $P_m(k) = [\hat{A}(k, q^{-1}), \hat{B}(k, q^{-1})]$; the estimation error is denoted by $e(k) = y(k) - \phi^T(k-d)\hat{\theta}(k)$.

Let $C(k)$ denote the adaptive controller, it may be designed by any existing principle, for example, pole-placement. To be specific, $C(k)$ is determined by a mapping from model parameter set to controller parameter set, i.e.

$$f : \mathbb{P} \rightarrow \mathbb{C} \tag{2.5}$$

$\mathbb{P}$ denotes the model parameter set including the true model of the plant. $\mathbb{C}$ is the corresponding controller parameter set. Then we have

$$C(k) = f(P_m(k)) \tag{2.6}$$

It is not difficult to imagine that if plant P satisfies some certain conditions, like controllability, there must exists a time-invariant controller C, which was named the "virtual" controller.

$$C = f(P) \tag{2.7}$$

Furthermore, let $\theta_c(k)$ and θ_c be the parametrizations of $C(k)$ and C respectively. The outputs of these two controllers are

$$u(k) = \phi_c^T(k)\theta_c(k) \tag{2.8}$$

$$u_0(k) = \phi_c^T(k)\theta_c \tag{2.9}$$

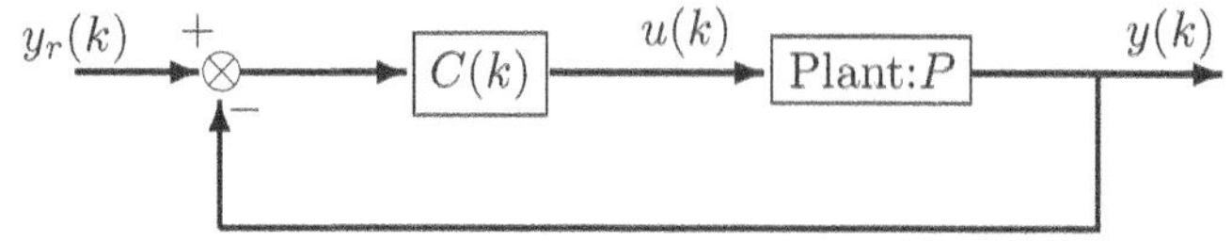

Fig. 2.1 Block diagram for deterministic STC system

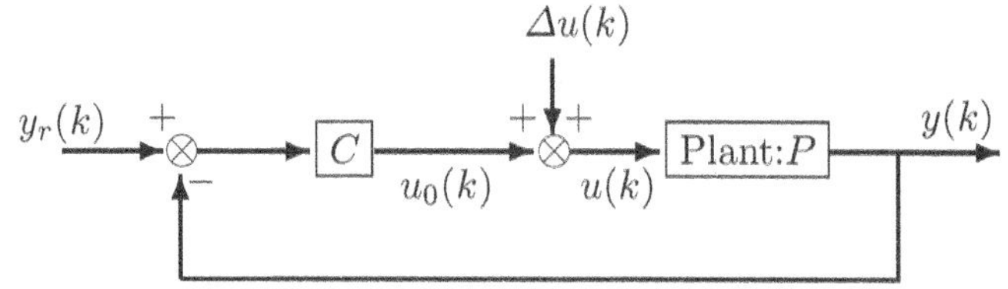

Fig. 2.2 Block diagram for deterministic VES I

where

$$\phi_c^T(k) = [y(k), y(k-1), \cdots, y(k-s_1), u(k-1), \cdots, u(k-s_2), y_r(k), \cdots, y_r(k-s_3)] \tag{2.10}$$

The specific elements of $\phi_c(k)$ and the limited integers s_1, s_2, s_3 depend on the detail control strategy, $y_r(k)$ is a given bounded reference signal, $y_r(k) = 0$ for $k < 0$. The block diagram of the STC system is shown in Fig. 2.1, in which the estimator details are omitted here for simplicity. Its VESs are shown in Figs. 2.2 and 2.3.

Figure 2.2 corresponds to the situation that parameter estimates converge to the true values; Fig. 2.3 corresponds to the situation that the parameter estimates converge, but are not consistent.

In Fig. 2.2, $\Delta u(k)$ is the output difference between controllers $C(k)$ and C under the same circumstances.

$$\Delta u(k) = u(k) - u_0(k) = \phi_c^T(k)[\theta_c(k) - \theta_c] \tag{2.11}$$

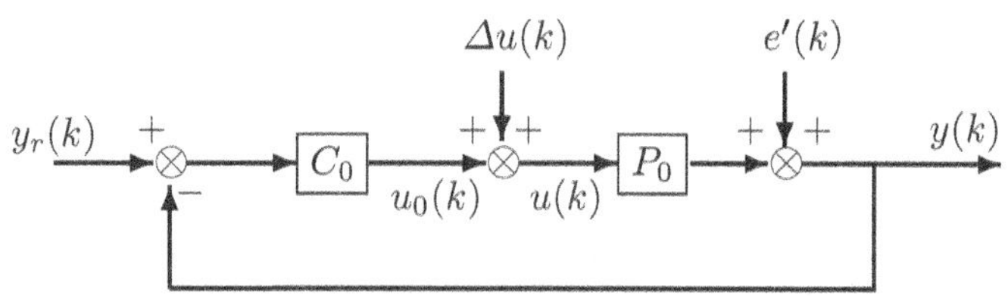

Fig. 2.3 Block diagram for deterministic VES II

Notice that $\Delta u(k)$ cannot be calculated from its definition, because the parameters of virtual controller C is unknown, but fortunately it can be estimated in other ways (details will be given later).

In Fig. 2.3, assume $P_m(k)$, i.e., $\hat{\theta}(k)$ converges to P_0, i.e., θ_0, then we have

$$C_0 = f(P_0) \tag{2.12}$$

And further we have

$$\begin{aligned} e'(k) &= y(k) - \phi^T(k-d)\theta_0 \\ &= y(k) - \phi^T(k-d)\hat{\theta}(k) + \phi^T(k-d)[\hat{\theta}(k) - \theta_0] \\ &= e(k) + \phi^T(k-d)[\hat{\theta}(k) - \theta_0] \end{aligned} \tag{2.13}$$

$$\Delta u(k) = u(k) - u_0(k) = \phi_c^T(k)[\theta_c(k) - \theta_{c0}] \tag{2.14}$$

where θ_{c0} is the parameterization of C_0.

2.2.2 *Main Results*

2.2.2.1 Parameter Estimates Converge to True Values

For this situation, we have the following result.

Theorem 2.1 *The STC system of a deterministic plant with known structure information (i.e., n, m, and d) is stable and convergent, if only the following properties are satisfied:*

(1) P is controllable and the control strategy is well defined such that $\Sigma\{C, P\}$ constitutes a stable closed-loop system;
(2) The parameter estimates converge to true values;
(3) The mapping from estimated parameters into controller parameters is continuous at θ.

Proof The VES (as shown in Fig. 2.2) can be decomposed into two subsystems, as shown in Figs. 2.4 and 2.5, respectively.

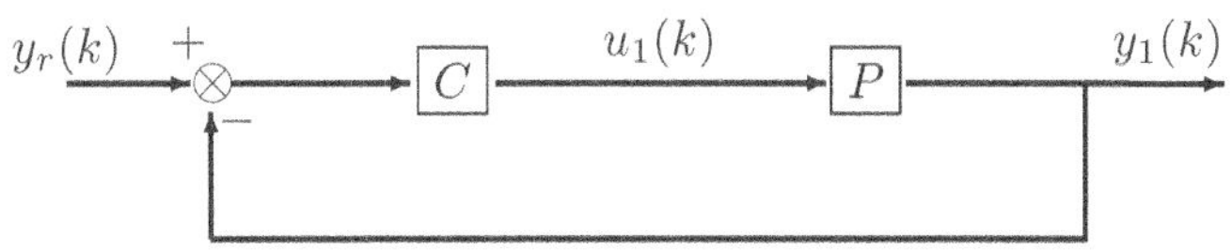

Fig. 2.4 Block diagram for subsystem 1 of deterministic VES I

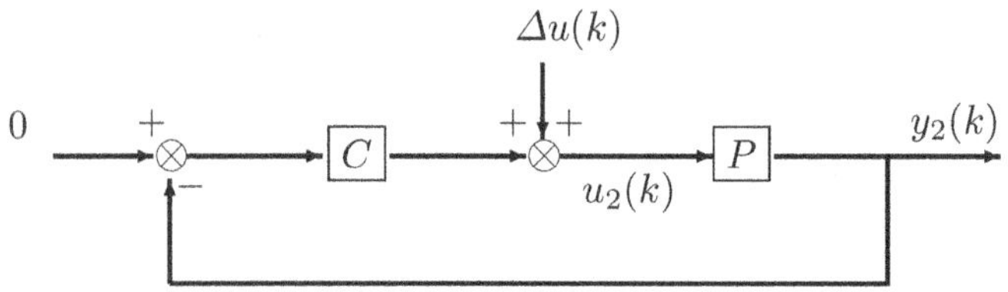

Fig. 2.5 Block diagram for subsystem 2 of deterministic VES I

According to superposition principle of linear system, we have

$$y(k) = y_1(k) + y_2(k), \quad u(k) = u_1(k) + u_2(k) \tag{2.15}$$

To simplify the proof, we need define a new vector $\tilde{\phi}(k)$, whose elements are the union of that of $\phi(k-d)$ and $\phi_c(k)$. Without loss of generality, assume $s_1 < n, s_2 < m$, then $\tilde{\phi}(k)$ takes the form of

$$\tilde{\phi}(k) = [y(k), \ldots, y(k-n), u(k-1), \ldots, u(k-d-m), y_r(k), \ldots, y_r(k-s_3)] \tag{2.16}$$

Similarly, define its counterparts $\tilde{\phi_1}(k)$ in subsystem1 (Fig. 2.4), and $\tilde{\phi_2}(k)$ in subsystem 2 (Fig. 2.5), respectively, i. e.

$$\tilde{\phi_1}(k) = [y_1(k), \ldots, y_1(k-n), u_1(k-1), \ldots, u_1(k-d-m), y_r(k), \ldots, y_r(k-s_3)] \tag{2.17}$$

$$\tilde{\phi_2}(k) = [y_2(k), \ldots, y_2(k-n), u_2(k-1), \ldots, u_2(k-d-m), 0, \ldots, 0] \tag{2.18}$$

Then, we obtain

$$\tilde{\phi}(k) = \tilde{\phi_1}(k) + \tilde{\phi_2}(k), \quad \|\phi(k-d)\| = O(\|\tilde{\phi}(k)\|), \quad \|\phi_c(k)\| = O(\|\tilde{\phi}(k)\|)$$

Further, from Lemma A.2 in Appendix A, we have the following fact that

$$\|\tilde{\phi}(k)\| = O(\|\tilde{\phi}(k-i)\|) + M, \quad 0 \le M < \infty \tag{2.19}$$

where i is a limited integer. From Condition (1), we know that subsystem 1 is stable. So that

$$|y_1(k)| < \infty, \quad |u_1(k)| < \infty, \quad \|\tilde{\phi_1}(k)\| < \infty \tag{2.20}$$

As for subsystem 2, by Condition (2) and Condition (3) it follows that

$$\Delta u(k) = o(\|\phi_c(k)\|) = o(\|\tilde{\phi}(k)\|) \tag{2.21}$$

Considering Condition (1) we know that subsystem 2 is also a stable closed-loop system. Then we get

$$|y_2(k)| = o(\|\tilde{\phi}(k)\|), \quad |u_2(k)| = o(\|\tilde{\phi}(k)\|) \tag{2.22}$$

This together with the definition of $\tilde{\phi}_2(k)$ and (2.22) yields

$$\|\tilde{\phi}_2(k)\| = o((\alpha + \|\tilde{\phi}(k-l)\|)), \quad 0 \le \alpha < \infty \tag{2.23}$$

where $l = \max(n, d+m)$ is a limited integer.

Then, by way of contradiction, we conclude that $\|\tilde{\phi}(k)\|$ is bounded. For details, see Lemma A.2. It means that the VES is stable, i.e.

$$|y(k)| < \infty, \quad |u(k)| < \infty, \quad \|\tilde{\phi}(k)\| < \infty \tag{2.24}$$

This together with (2.22) leads to

$$|y_2(k)| = o(1), \quad |u_2(k)| = o(1) \tag{2.25}$$

Consequently, from (2.15) we have

$$y(k) \to y_1(k), \quad u(k) \to u_1(k) \tag{2.26}$$

Then we know the VES is convergent.

2.2.2.2 Parameter Estimates Converge But Are Not Consistent

We have the following result for this situation.

Theorem 2.2 *The STC system of a deterministic plant with known structure information (i.e., n, m, and d) is stable and convergent, if only the following properties are satisfied:*

(1) *$P_m(k)$ is uniformly controllable and converges to P_0 with*

$$e(k) = y(k) - \phi^T(k-d)\hat{\theta}(k) = o(\alpha + \|\phi(k-d)\|), 0 < \alpha < \infty \tag{2.27}$$

(2) *The control strategy is well defined such that $\Sigma\{C_0, P_0\}$ constitutes a stable closed-loop system;*

(3) *The mapping from estimated parameters into controller parameters is continuous at θ_0.*

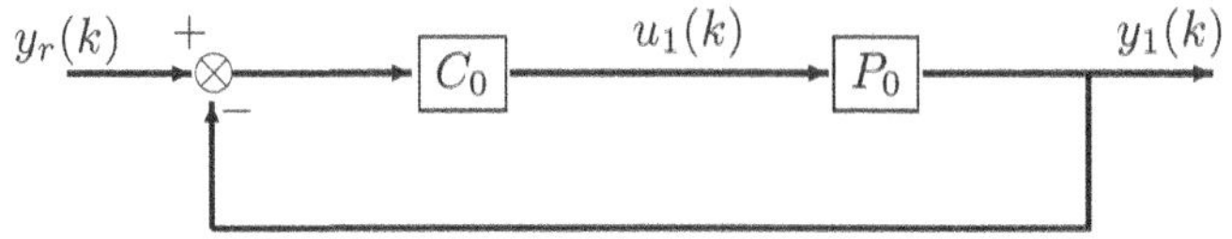

Fig. 2.6 Block diagram for subsystem 1 of deterministic VES II

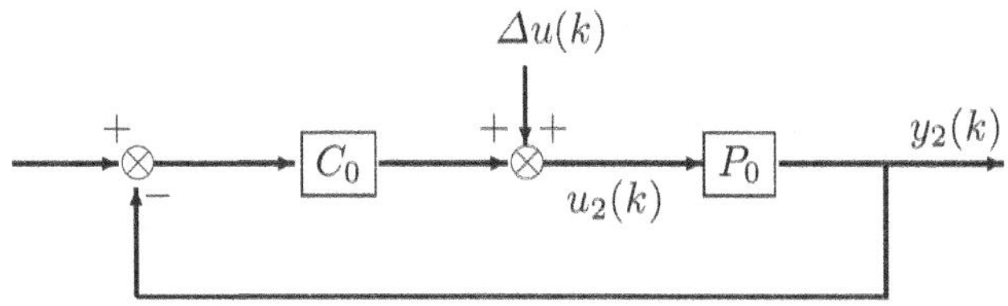

Fig. 2.7 Block diagram for subsystem 2 of deterministic VES II

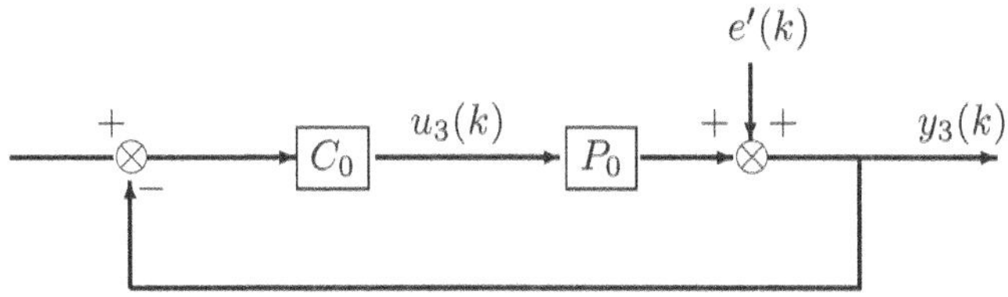

Fig. 2.8 Block diagram for subsystem 3 of deterministic VES II

Remark 2.1 For the definition of uniform controllability, the reader is referred to Ref. [24]. It is needed to guarantee the existence of $u(k)$.

Proof Decompose the corresponding VES (as shown in Fig. 2.3) into three subsystems, as shown in Figs. 2.6, 2.7, and 2.8, respectively.

According to superposition principle we have

$$y(k) = y_1(k) + y_2(k) + y_3(k), \quad u(k) = u_1(k) + u_2(k) + u_3(k) \tag{2.28}$$

Similar to Sect. 2.2.2.1, define $\tilde{\phi}(k)$, $\tilde{\phi}_1(k)$, $\tilde{\phi}_2(k)$, and $\tilde{\phi}_3(k)$. Then we get

$$\tilde{\phi}(k) = \tilde{\phi}_1(k) + \tilde{\phi}_2(k) + \tilde{\phi}_3(k), \quad \|\phi(k-d)\| = O(\|\tilde{\phi}(k)\|), \quad \|\phi_c(k)\| = O(\|\tilde{\phi}(k)\|)$$

First, we know that Fig. 2.6 represents a stable deterministic close-loop control system under Condition (2). That means

$$|u_1(k)| < \infty, \quad |y_1(k)| < \infty, \quad \|\tilde{\phi}_1(k)\| < \infty \tag{2.29}$$

Second, in subsystem 2 as shown in Fig. 2.7, Conditions (1) and (3) guarantee

$$\Delta u(k) = o(\|\phi_c(k)\|) = o(\|\tilde{\phi}(k)\|) \tag{2.30}$$

Once again by Condition (2), subsystem 2 is stable, then we get

$$|y_2(k)| = o(\|\tilde{\phi}(k)\|), \quad |u_2(k)| = o(\|\tilde{\phi}(k)\|), \quad \|\tilde{\phi}_2(k)\| = o(\|\tilde{\phi}(k-l)\|) \tag{2.31}$$

where $l = \max(n, d+m)$ is a limited integer.

Finally, we consider subsystem 3 as shown in Fig. 2.8. By Condition (1) and Eq. (2.13), we have

$$e'(k) = o(\alpha + \|\phi(k-d)\|) = o(\alpha + \|\tilde{\phi}(k)\|) \tag{2.32}$$

Also by Condition (2), we know that subsystem 3 is stable, then means

$$|y_3(k)| = o(\alpha + \|\tilde{\phi}(k)\|), \quad |u_3(k)| = o(\alpha + \|\tilde{\phi}(k)\|), \quad \|\tilde{\phi}_3(k)\| = o(\alpha + \|\tilde{\phi}(k-l)\|) \tag{2.33}$$

Then we come to the result of Theorem 2.2 following the similar procedures of the proof of Theorem 2.1. □

2.2.2.3 Parameter Estimates May Not Converge

First, we consider one-step-ahead control strategy combined with arbitrary parameter estimation algorithm. The reason we pick it out of the general case is that we want to explain why such kind of STC system can achieve stability and convergence without requiring the convergence of parameter estimates. In this situation we still use VES as shown in Fig. 2.2.

Second, we consider a more general case: arbitrary control strategy (such as pole-placement) combined with arbitrary parameter estimation algorithm. The VES for this situation will take a new form, i.e., 'slow switching' linear deterministic system.

Theorem 2.3 *The STC system of a deterministic plant with known structure information (i.e., n, m, and d) is stable and convergent, if only the following properties are satisfied:*

(1) $B(q^{-1})$ *is a Hurwitz polynomial with* $b_0 \neq 0$;
(2) $\hat{b}_0 \neq 0$ *or some modification on* $\hat{b}_0$ *to guarantee the existence of* $u(k)$;
(3) Parameter estimates yield

$$e(k) = y(k) - \phi^T(k-d)\hat{\theta}(k) = o(\alpha + \|\phi(k-d)\|), 0 < \alpha < \infty \tag{2.34}$$

Proof The key point of the above proof procedures is the property of $\Delta u(k)$ in Fig. 2.2 or the properties of $\Delta u(k)$ and $e(k)$ in Fig. 2.3. In Theorems 2.1 and 2.2, the convergence of parameter estimation is used to obtain the properties of $\Delta u(k)$ and $e(k)$. In the following, we will show that for one-step-ahead control strategy, the convergence of parameter estimation is not necessary to keep the property of $\Delta u(k)$ in its VES (Fig. 2.2). It is the unique characteristic of one-step-ahead control system.

Rewrite Eq. (2.1) in prediction form

$$y(k+d) = \phi^T(k)\hat{\theta}(k) \tag{2.35}$$

Then one-step-ahead adaptive control signal $u(k)$ is decided by

$$\phi^T(k)\hat{\theta}(k) = y_r(k+d) \tag{2.36}$$

This means

$$\begin{aligned}\hat{\theta}_{n+1}u(k) = &-\hat{\theta}_1 y(k) - \cdots - \hat{\theta}_n y(k-n+1) - \hat{\theta}_{n+2}u(k-1) - \cdots \\ &- \hat{\theta}_{n+m+d}u(k-m-d+1) + y_r(k+d)\end{aligned} \tag{2.37}$$

Accordingly

$$\begin{aligned}\theta_{n+1}u_0(k) = &-\theta_1 y(k) - \cdots - \theta_n y(k-n+1) - \theta_{n+2}u(k-1) \\ &- \cdots - \theta_{n+m+d}u(k-m-d+1) + y_r(k+d)\end{aligned} \tag{2.38}$$

It is clear that

$$\begin{aligned}\theta_{n+1}u_0(k) - \hat{\theta}_{n+1}u(k) &= \theta_{n+1}u_0(k) - \theta_{n+1}u(k) + (\theta_{n+1} - \hat{\theta}_{n+1})u(k) \\ &= \theta_{n+1}\Delta u(k) + (\theta_{n+1} - \hat{\theta}_{n+1})u(k)\end{aligned} \tag{2.39}$$

Then, we get

$$\Delta u(k) = \frac{1}{\theta_{n+1}}[\theta_{n+1}u_0(k) - \hat{\theta}_{n+1}u(k) + (\hat{\theta}_{n+1} - \theta_{n+1})u(k)] \tag{2.40}$$

This together with (2.37) and (2.38) gives

$$\begin{aligned}\Delta u(k) &= \frac{1}{\theta_{n+1}}[\phi^T(k-d)\theta - \phi^T(k-d)\hat{\theta}(k)] \\ &= \frac{1}{\theta_{n+1}}[y(k) - \phi^T(k-d)\hat{\theta}(k)] \\ &= O(|e(k)|)\end{aligned} \tag{2.41}$$

Here, θ_{n+1} is actually b_0 in Eq. (2.1) when $d = 1$. Based on Condition (3) and Eq. (2.41), we have

$$\Delta u(k) = o(\alpha + \|\phi(k-d)\|), 0 < \alpha < \infty \tag{2.42}$$

The rest procedures of the proof are similar to that of Theorem 2.1: Decompose the VES (Fig. 2.2) into two subsystems as shown in Fig. 2.4 and Fig. 2.5 respectively; and then by way of contradiction, it is straightforward to get to the conclusion of stability and convergence. □

Remark 2.2 Generally speaking, most deterministic parameter estimation algorithms, like projection algorithm or gradient algorithm, can guarantee

$$y(k) - \phi^T(k-d)\hat{\theta}(k-1) = o(\alpha + \|\phi(k-d)\|)$$

$$\lim_{k\to\infty} \|\hat{\theta}(k) - \hat{\theta}(k-l)\| \to 0, \quad l \text{ is a finite integer.}$$

Then we know that Condition (3) of Theorem 2.3, i.e., Eq. (2.34) holds.

Next, we address the general case: arbitrary control strategy combined with arbitrary parameter estimation algorithm.

Theorem 2.4 *The STC system of deterministic plant with known structure information (i.e., n, m, and d) is stable and convergent, if only the following properties are satisfied:*

(1) Control strategy is well defined to stabilize parameter known plant and tracking the bounded reference signal $y_r(k)$*;*
(2) $\|\hat{\theta}(k)\| \le M < \infty$ *, and* $\|\hat{\theta}(k) - \hat{\theta}(k-l)\| \to 0$ *,* l *is a finite integer;*
(3) The parameter estimates error satisfies

$$e(k) = y(k) - \phi^T(k-d)\hat{\theta}(k) = o(\alpha + \|\phi(k-d)\|), 0 < \alpha < \infty \tag{2.43}$$

(4) The mapping from model parameters into controller parameters is continuous in $\hat{\theta}(k)$*, for* $\|\hat{\theta}(k)\| \le M < \infty$*.*

Remark 2.3 According to Cauchy convergence criterion, condition (2), i. e, $\|\hat{\theta}(k)\| < \infty$, $\|\hat{\theta}(k) - \hat{\theta}(k-l)\| \to 0$, l is a finite integer, does not mean the convergence of $\hat{\theta}(k)$. In fact, $\|\hat{\theta}(k)\| < \infty$, and $\|\hat{\theta}(k) - \hat{\theta}(k-l)\| \to 0$, $l \to \infty$ means the convergence of $\hat{\theta}(k)$. Here in Theorem 2.4, 'continuous' is identical to 'uniformly continuous', because the domain $\|\hat{\theta}(k)\| \le M < \infty$ is closed(compact).

Proof An STC system (as shown in Fig. 2.1) is obviously equivalent to the system as shown in Fig. 2.9 in the input-output sense. Further, we obtain an equivalent system

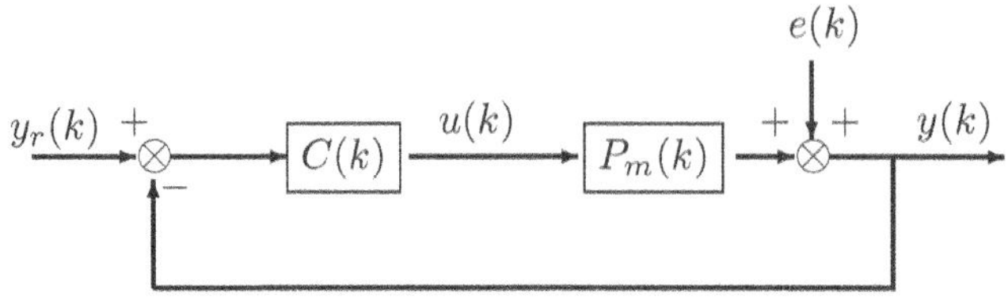

Fig. 2.9 Block diagram for deterministic VES III

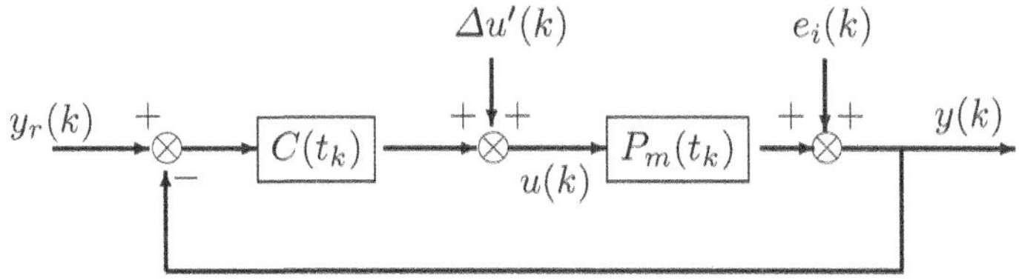

Fig. 2.10 Block diagram for deterministic VES IV

(as shown in Fig. 2.10), which is a slow switching system under the Condition 2), where

$$\Delta u'(k) = \phi_c^T(k)[\hat{\theta}_c(k) - \hat{\theta}_c(t_k)]$$

The sequence of t_k satisfies

$$t_{k+i} = \begin{cases} k+i & \text{if } i = T \\ t_k & \text{if } i < T \end{cases} \qquad T = \begin{cases} \tau_D + 1 & \text{if } k \geq K \\ 1 & \text{if } k < K \end{cases} \tag{2.44}$$

where $t_0 = 0$, i takes values from 1 to T, once $i = T$ then i return to 1. where K is a sufficient large integer, τ_D is the so-called "dwell time" to guarantee the stability of a slow switching control system, which, by Condition (2), is a constant depending on the upper and lower bounds of $\hat{\theta}_k$ [37–39].

In Fig. 2.10, we have

$$e_i(k) = y(k) - \phi^T(k-d)\hat{\theta}_{t_k} = y(k) - \phi^T(k-d)\hat{\theta}(k) + \phi^T(k-d)[\hat{\theta}_k - \hat{\theta}_{t_k}] \tag{2.45}$$

Conditions (2) and (3) lead to

$$e_i(k) = o(\alpha + \|\phi(k-d)\|), 0 < \alpha < \infty \tag{2.46}$$

Conditions (2) and (4) guarantee the property of $\Delta u'(k)$

$$\Delta u'(k) = o(\|\phi_c(k)\|) \tag{2.47}$$

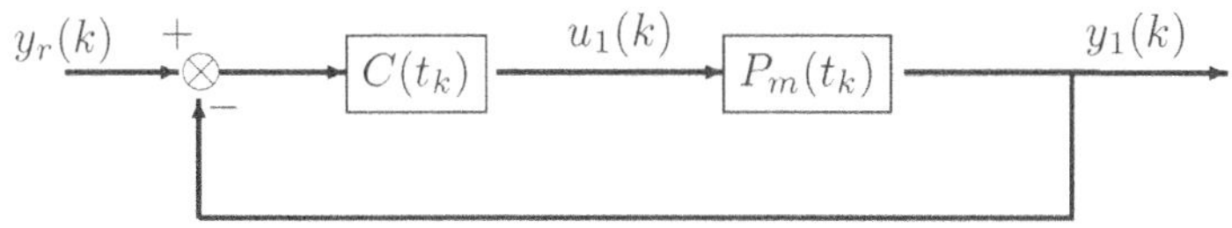

Fig. 2.11 Block diagram for subsystem 1 of deterministic VES IV

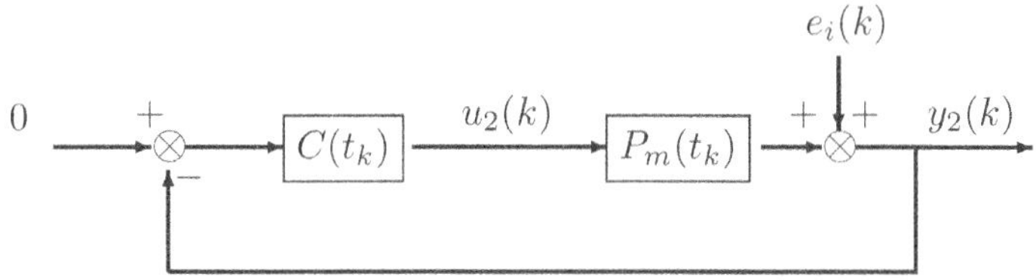

Fig. 2.12 Block diagram for subsystem 2 of deterministic VES IV

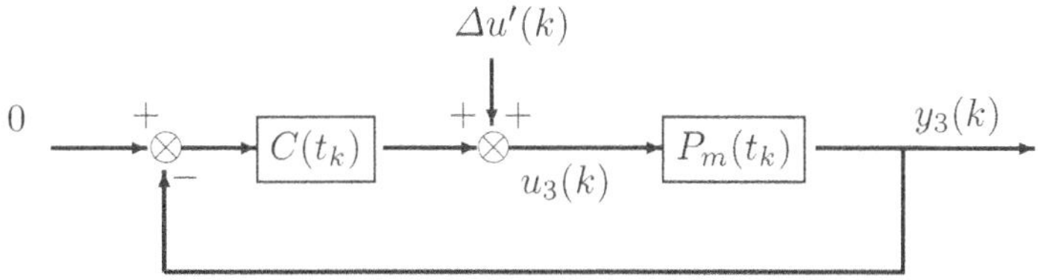

Fig. 2.13 Subsystem 3 of deterministic VES IV

Next, according to Lemma A.3 in Appendix A, the VES IV can be decomposed into three subsystems, as shown in Figs. 2.11, 2.12, and 2.13, respectively. They are all deterministic slow switching systems.

By superposition principle, we get

$$y(k) = y_1(k) + y_2(k) + y_3(k) \tag{2.48}$$

$$u(k) = u_1(k) + u_2(k) + u_3(k) \tag{2.49}$$

Similar to Sects. 2.2.2.1 and 2.2.2.2, define $\tilde{\phi}(k)$, $\tilde{\phi}_1(k)$, $\tilde{\phi}_2(k)$ and $\tilde{\phi}_3(k)$.

Then, we have

$$\tilde{\phi}(k) = \tilde{\phi}_1(k) + \tilde{\phi}_2(k) + \tilde{\phi}_3(k)\|, \quad \phi(k-d)\| = O(\|\tilde{\phi}(k)\|), \quad \|\phi_c(k)\| = O(\|\tilde{\phi}(k)\|) \tag{2.50}$$

Note that in these subsystems, k is sufficiently large, i.e., $k \geq K$, and all the performances before K are counted into the initial conditions of the related systems. Then all the three subsystems are deterministic "slow switching" systems with large enough dwell time. Further, according to Condition (1) we know that for each given t_k, $C(t_k)$ and $P_m(t_k)$ constitute a stable closed-loop system. Thus, by Multiple

Lyapunov Function (MLF) methodology, subsystem 1 (as shown in Fig. 2.11) is stable [37–40]. That means

$$|u_1(k)| < \infty, \quad |y_1(k)| < \infty, \quad \|\tilde{\phi}_1(k)\| < \infty, \quad y(k) \to y_r(k) \tag{2.51}$$

Obviously, Subsystem 2 (Fig. 2.12) and Subsystem 3 (Fig. 2.13) are also stable, and their input signals were described by Eqs.(2.46) and (2.47) respectively. Then, taking account of (2.50), we obtain

$$|y_2(k)| = o(\alpha + \|\tilde{\phi}(k)\|), \quad |u_2(k)| = o(\alpha + \|\tilde{\phi}(k)\|), \quad \|\tilde{\phi}_2(k)\| = o(\alpha + \|\tilde{\phi}(k-l)\|) \tag{2.52}$$

and

$$|y_3(k)| = o(\|\phi_c(k)\|) = o(\|\tilde{\phi}(k)\|), \quad |u_3(k)| = o(\|\tilde{\phi}(k)\|), \quad \|\tilde{\phi}_3(k)\| = o(\|\tilde{\phi}(k-l)\|) \tag{2.53}$$

Then along the similar way of the proof of Theorem 2.1, we know that the STC system is stable, i.e., $\|\tilde{\phi}(k)\| < \infty$.

Next, we proceed to prove the convergence of the STC system. To reach this goal, we need first to prove the tracking performance of subsystem 1. We will make further use of MLF, but we need to change the angle of view a bit. Denote the tracking error of each "frozen" system $\sum\{C(\bar{t}_k), P_m(\bar{t}_k)\}$ by

$$e_{\bar{t}_k}(k) = y_{\bar{t}_k}(k) - y_r(k)$$

where $\bar{t}_k$ represents the "frozen" t_k.

Similarly, the tracking error of $\sum\{C(t_k), P_m(t_k)\}$, i.e., subsystem 1 (Fig. 2.11), is denoted by

$$e_{t_k}(k) = y_{t_k}(k) - y_r(k) = y_1(k) - y_r(k)$$

Then, we know that for each "frozen" system $\sum\{C(\bar{t}_k), P_m(\bar{t}_k)\}$, there exists a dynamic system

$$e_{\bar{t}_k}(k+1) = f_{\bar{t}_k}(e_{\bar{t}_k}(k), k)$$

Thus, for $\sum\{C(t_k), P_m(t_k)\}$, there exists a "slow switching" dynamic system

$$e_{t_k}(k+1) = f_{t_k}(e_{t_k}(k), k)$$

According to Condition (1), for each $\bar{t}_k$, the stable equilibrium is $e_{\bar{t}_k} = 0$, thus, there must exists a Lyapunov function $V_{\bar{t}_k}(e)$ satisfying

$$V_{\bar{t}_k}(e_{\bar{t}_k}(k)) \geq 0, \quad \Delta V_{\bar{t}_k}(e_{\bar{t}_k}(k)) < 0, \quad \lim_{k\to\infty} V_{\bar{t}_k}(e_{\bar{t}_k}(k)) = 0$$

Further, based on MLF methodology and "slow switching" property, we get similar result for subsystem 1, i.e.

$$V(e_{t_k}(k)) \geq 0, \quad \Delta V(e_{t_k}(k)) < 0, \quad \lim_{k\to\infty} V(e_{t_k}(k)) = 0$$

That implies the tracking performance of the "slow switching" subsystem 1, i.e.

$$e_{t_k}(k) = y_{t_k}(k) - y_r(k) = y_1(k) - y_r(k) \to 0$$

Then from (2.48), (2.49), (2.52), (2.53), and the fact $\|\tilde{\phi}(k)\| < \infty$, we know that the STC system is convergent. □

Remark 2.4 Instead of convergence of parameter estimates, Theorem 2.4 actually only requires an asymptotic property of parameter estimates, which is usually satisfied by most existing estimates algorithms under some mild assumptions.

Next, we extend the above result to the following linear time varying (LTV) system.

$$A(\gamma, q^{-1})y(k) = q^{-d}B(\gamma, q^{-1})u(k) \tag{2.54}$$

where

$$A(\gamma, q^{-1}) - 1 + a_1(\gamma)q^{-1} + \cdots + a_n(\gamma)q^{-n}$$
$$B(\gamma, q^{-1}) = b_0(\gamma) + b_1(\gamma)q^{-1} + \cdots + b_m(\gamma)q^{-m}$$

Denote

$$\theta^T(\gamma) = [a_1(\gamma), \ldots a_n(\gamma), b_0(\gamma), b_1(\gamma), b_m(\gamma)]$$

We need the following assumptions:

(A1) $\theta(\gamma)$ belongs to a compact subset of a finite-dimensional linear vector space, i.e. $\theta(\gamma) \in \Theta \subset \mathbb{R}^{n+m+1}$.
(A2) $a_i(\gamma), i = 1, \ldots, n$ and $b_j(\gamma), j = 0, 1, \ldots, m$ are all piecewise constants; the interval between consecutive parameter jumps is large enough.
(A3) The estimation algorithm can detect the parameters jump, say, by measuring the jump of estimation error $e(k) = y(k) - \phi^T(k-d)\hat{\theta}(k)$, and then the estimator can be reinitialized.

Then we have the following result.

Theorem 2.5 *Assume (A1), (A2) and (A3) hold. If the STC system of plant (2.54) with known n, m, and d, has the following properties:*

(1) *Control strategy is well defined to stabilize parameter known plant and tracking the bounded reference signal* $y_r(k)$;
(2) *For each "frozen"* γ *or* $\theta(\gamma)$, $\|\hat{\theta}(k)\| \leq M < \infty$, *and* $\|\hat{\theta}(k) - \hat{\theta}(k-l)\| \to 0$, l *is a finite integer.*
(3) *For each "frozen"* γ *or* $\theta(\gamma)$, *the estimation error satisfies*

$$e(k) = y(k) - \phi^T(k-d)\hat{\theta}(k) = o(\alpha + \|\phi(k-d)\|), 0 < \alpha < \infty$$

(4) *The mapping from model parameters into controller parameters is continuous in* $\hat{\theta}(k)$, *for* $\|\hat{\theta}(k)\| \leq M < \infty$.
Then the STC system is stable.

Proof Based on Theorem 2.4 and Conditions (1) to (4), we know that for each "frozen" γ or $\theta(\gamma)$, there exists a stable STC system denoted by $\sum\{C(k), P_{\theta(\gamma)}\}$. Further, Assumptions (A2) and (A3) guarantee the "slow switching" between these stable STC systems with large enough dwell time. Then, by MLF methodology plus Assumption (A1), the overall STC system $\sum\{C(k), P_{\Theta}\}$ is stable.

The proof is completed. □

2.3 Stochastic Self-Tuning Control System

For simplicity of description, we will first consider stochastic plant P with white noise, and finally we will extend the results to plant P with colored noise. And all the limit operations in this section are in the sense of probability one.

2.3.1 Construction of Virtual Equivalent System

Similar to deterministic STC, three kinds of VES will be used for stochastic STC. In this subsection we will present two of them, which are comparatively easy to understand. As for the third situation that parameter estimates may not converge, we will use another type of VES which is a stochastic slow switching system. It will be described later in Sect. 2.3.2.

Consider

$$A(q^{-1})y(k) = q^{-d}B(q^{-1})u(k) + \omega(k) \tag{2.55}$$

where

$$A(q^{-1}) = 1 + a_1q^{-1} + \cdots + a_nq^{-n}$$
$$B(q^{-1}) = b_0 + b_1q^{-1} + \cdots + b_mq^{-m}$$

where $y(k)$, $u(k)$ and $\omega(k)$ are the output, input, and disturbance, respectively, $y(k) = 0$, $u(k) = 0$, $\omega(k) = 0$ for $k < 0$, $A(q^{-1})$ and $B(q^{-1})$ are polynomials in backward-shift operator q^{-1} with unknown coefficients and with known orders or upper bounds of orders n, m.

The plant equation can be rewritten as

$$y(k) = \phi^T(k-d)\theta + \omega(k) \tag{2.56}$$

$$\phi^T(k-d) = [y(k-1), \ldots, y(k-n), u(k-d), \ldots, u(k-d-m)] \tag{2.57}$$

$$\theta^T = [-a_1, \ldots, -a_n, b_0, b_1, \ldots, b_m] \tag{2.58}$$

We use $P_m(k)$ and $\hat{\theta}(k)$ to denote the estimated model and estimated parameters respectively.

Let $C(k)$ denote the adaptive controller, it may be designed by any existing principle, such as pole placement. To be specific, $C(k)$ is determined by a mapping from model parameters into controller parameters, i.e.

$$f : \mathbb{P} \to \mathbb{C} \tag{2.59}$$

where $\mathbb{P}$ is the set of estimated models including the true model of the plant. $\mathbb{C}$ is the corresponding controllers set. Then we denote

$$C(k) = f(P_m(k)) \tag{2.60}$$

It is not difficult to imagine that if plant P satisfies some conditions, like controllability, there must exists a time-invariant controller, which was the so called "virtual controller". That is

$$C = f(P) \tag{2.61}$$

Furthermore, let $\theta_c(k)$ and θ_c be the parameterizations of $C(k)$, and C, respectively. These two controllers generate outputs

$$u(k) = \phi_c^T(k)\theta_c(k) \tag{2.62}$$

$$u_0(k) = \phi_c^T(k)\theta_c \tag{2.63}$$

where

$$\phi_c^T(k) = [y(k), y(k-1), \ldots, y(k-s_1), u(k-1), \ldots, u(k-s_2), y_r(k), \ldots, y_r(k-s_3)] \tag{2.64}$$

The elements of $\phi_c(k)$ and the limited integers s_1, s_2, s_3 depend on specific control strategy, $y_r(k)$ is a given bounded reference signal of the system, $y_r(k) = 0$ for $k < 0$. The STC system is shown in Fig. 2.14 (the estimator is omitted for simplicity), its virtual equivalent systems are shown in Figs. 2.15 and 2.16.

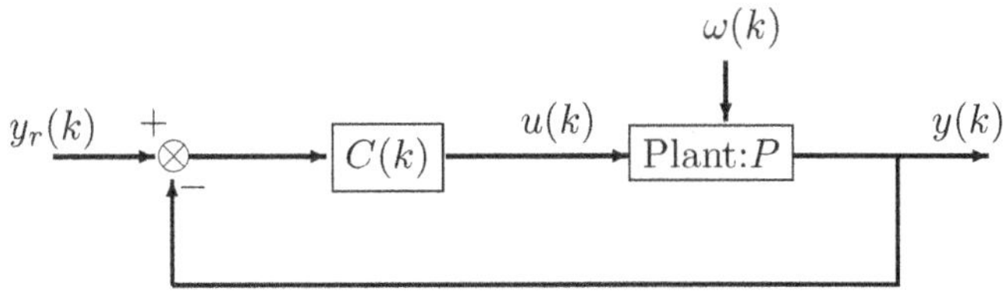

Fig. 2.14 Block diagram for stochastic self-tuning Control System

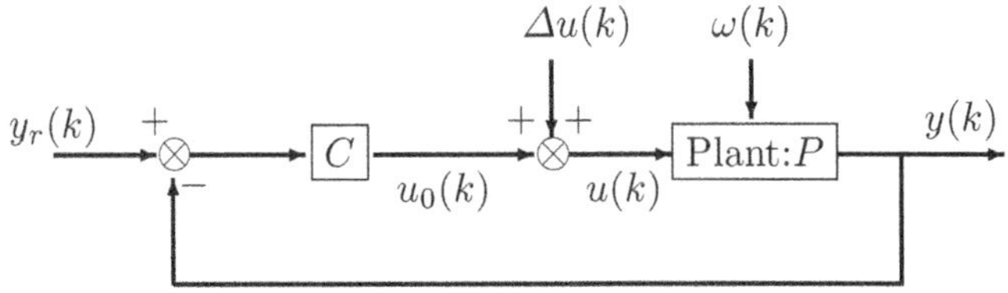

Fig. 2.15 Block diagram for stochastic VES I

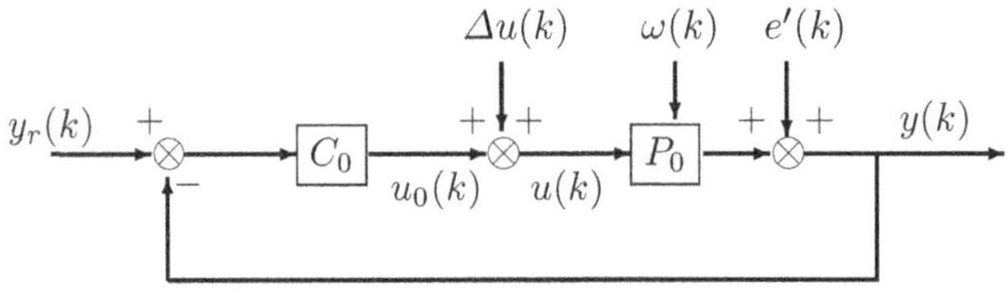

Fig. 2.16 Block diagram for stochastic VES II

Figure 2.15 corresponds to the situation that the parameter estimates converge to the true values; Fig. 2.16 corresponds to the situation that the parameter estimates converge, but are not consistent.

In Fig. 2.15, $\Delta u(k)$ is the difference between outputs of controllers $C(k)$ and C.

$$\Delta u(k) = u(k) - u_0(k) = \phi_c^T(k)[\theta_c(k) - \theta_c] \tag{2.65}$$

Notice that $\Delta u(k)$ cannot be calculated from its definition, because the parameter vector θ_c of virtual controller C is unknown, but it can be estimated in other ways (for details, see the proofs of theorems later on).

In Fig. 2.16, we assume $P_m(k)$, i.e., $\hat{\theta}(k)$ converges to P_0, i.e., θ_0, then we get

$$C_0 = f(P_0) \tag{2.66}$$

And that

$$\begin{aligned} e'(k) &= y(k) - \phi^T(k-d)\theta_0 - \omega(k) \\ &= \phi^T(k-d)[\theta - \hat{\theta}(k)] + \phi^T(k-d)[\hat{\theta}(k) - \theta_0] \end{aligned} \tag{2.67}$$

$$\Delta u(k) = u(k) - u_0(k) = \phi_c^T(k)[\theta_c(k) - \theta_{c0}] \tag{2.68}$$

where θ_{c0} is the parameterization of C_0.

2.3.2 *Main Results*

2.3.2.1 Parameter Estimates Converge to True Values

Consider a stochastic STC system consisted of arbitrary control strategy and arbitrary estimates algorithm. We have the following result.

Theorem 2.6 *The STC system of a stochastic plant with known structure information (i.e., n, m, and d) is stable and convergent, if only the following properties are satisfied:*

(1) P is controllable and the control strategy is well defined such that $\Sigma\{C, P\}$ constitutes a stable closed-loop system;
(2) The parameter estimates converge to the true values;
(3) The mapping from estimated parameters into controller parameters is continuous at θ.

Proof Since the VES as shown in Fig. 2.15 is linear and time-invariant in structure. Then it can be decomposed into two subsystems, as shown in Figs. 2.17, and 2.18, respectively.

By superposition principle of linear system, we obtain

$$y(k) = y_1(k) + y_2(k), \quad u(k) = u_1(k) + u_2(k) \tag{2.69}$$

Similar to the deterministic STC, we define $\widetilde{\phi}(k)$, $\widetilde{\phi}_1(k)$, and $\widetilde{\phi}_2(k)$, respectively. Then we get

$$\widetilde{\phi}(k) = \widetilde{\phi}_1(k) + \widetilde{\phi}_2(k), \quad \|\phi_c(k)\| = O(\|\widetilde{\phi}(k)\|) \tag{2.70}$$

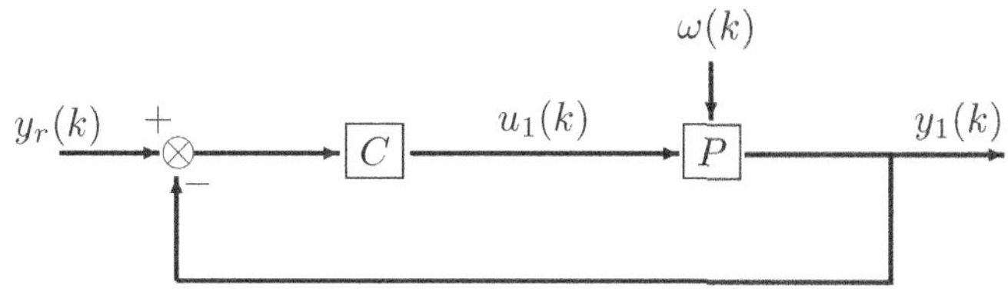

Fig. 2.17 Block diagram for subsystem 1 of stochastic VES I

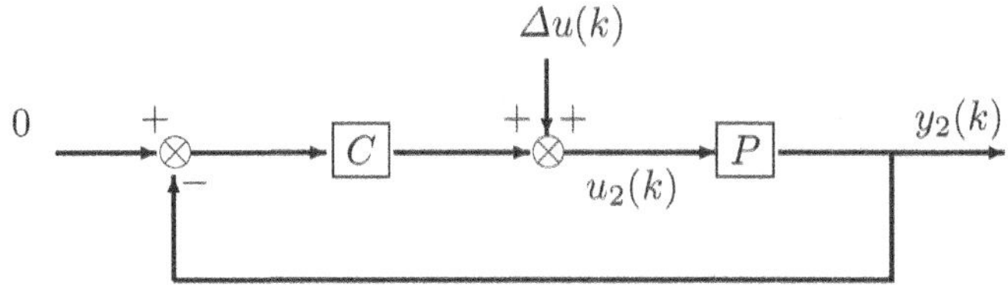

Fig. 2.18 Block diagram for subsystem 2 of stochastic VES I

By Condition (1), subsystem 1 is a stable stochastic system. That means

$$\frac{1}{n}\sum_{k=1}^{n}\|\widetilde{\phi}_1(k)\|^2 < \infty, \quad \frac{1}{n}\sum_{k=1}^{n}|y_1(k) - y_r(k)|^2 < \infty \tag{2.71}$$

As for subsystem 2, it is also a stable closed-loop system, and that it is a deterministic system. Then by Conditions (2) and (3) we know that

$$\Delta u(k) = o(\|\phi_c(k)\|) = o(\|\widetilde{\phi}(k)\|) \tag{2.72}$$

$$\frac{1}{n}\sum_{k=1}^{n}|\Delta u(k)|^2 = o\Big(\frac{1}{n}\sum_{k=1}^{n}\|\widetilde{\phi}(k)\|^2\Big) \tag{2.73}$$

Thus, it follows that

$$\frac{1}{n}\sum_{k=1}^{n}|y_2(k)|^2 = o\Big(\frac{1}{n}\sum_{k=1}^{n}\|\widetilde{\phi}(k)\|^2\Big) \tag{2.74}$$

$$\frac{1}{n}\sum_{k=1}^{n}|u_2(k)|^2 = o\Big(\frac{1}{n}\sum_{k=1}^{n}\|\widetilde{\phi}(k)\|^2\Big) \tag{2.75}$$

Equations (2.72) and (2.73) imply that

$$\frac{1}{n}\sum_{k=1}^{n}\|\widetilde{\phi}_2(k)\|^2 = o\Big(\frac{1}{n}\sum_{k=1}^{n}\|\widetilde{\phi}(k)\|^2\Big) \tag{2.76}$$

Making use of the Cauchy inequality and the triangle inequality, we have

$$\begin{aligned}\tfrac{1}{n}\sum_{k=1}^{n}\|\widetilde{\phi}(k)\|^2 &= \tfrac{1}{n}\sum_{k=1}^{n}\|\widetilde{\phi}_1(k) + \widetilde{\phi}_2(k)\|^2 \\ &\le \tfrac{1}{n}\sum_{k=1}^{n}\|\widetilde{\phi}_1(k)\|^2 + \tfrac{1}{n}\sum_{k=1}^{n}\|\widetilde{\phi}_2(k)\|^2 + \tfrac{2}{n}\sum_{k=1}^{n}\|\widetilde{\phi}_1(k)\|\sum_{k=1}^{n}\|\widetilde{\phi}_2(k)\|\end{aligned} \tag{2.77}$$

Suppose, by contradiction, that $\frac{1}{n}\sum_{k=1}^{n}\|\widetilde{\phi}(k)\|^2$ is unbounded, there must exists a subsequence $\frac{1}{n}\sum_{k=1}^{n}\|\widetilde{\phi}(p_k)\|^2 \to \infty$, satisfying

$$\begin{aligned}\frac{1}{n}\sum_{k=1}^{n}\|\widetilde{\phi}(p_k)\|^2 =& \frac{1}{n}\sum_{k=1}^{n}\|\widetilde{\phi}_1(p_k)+\widetilde{\phi}_2(p_k)\|^2\\ \leq& \frac{1}{n}\sum_{k=1}^{n}\|\widetilde{\phi}_1(p_k)\|^2+\frac{1}{n}\sum_{k=1}^{n}\|\widetilde{\phi}_2(p_k)\|^2+\frac{2}{n}\sum_{k=1}^{n}\|\widetilde{\phi}_1(p_k)\|\sum_{k=1}^{n}\|\widetilde{\phi}_2(p_k)\|\end{aligned} \tag{2.78}$$

Besides, by the fact that the limits of sequence and its subsequences are the same, Eq. (2.74) indicates

$$\frac{1}{n}\sum_{k=1}^{n}\|\widetilde{\phi}_2(p_k)\|^2 = o\left(\frac{1}{n}\sum_{k=1}^{n}\|\widetilde{\phi}(p_k)\|^2\right) \tag{2.79}$$

This together with (2.76) and the fact $\sum_{k=1}^{n}\|\widetilde{\phi}_1(p_k)\|^2 \leq \sum_{k=1}^{n}\|\widetilde{\phi}_1(k)\|^2 < \infty$, yields, by squeezing, an absurd conclusion that

$$\frac{\sum_{k=1}^{n}\|\widetilde{\phi}(p_k)\|^2}{\sum_{k=1}^{n}\|\widetilde{\phi}(p_k)\|^2} \to 0$$

Then we conclude that

$$\frac{1}{n}\sum_{k=1}^{n}\|\widetilde{\phi}(k)\|^2 < \infty \tag{2.80}$$

That means the stability of the STC system. Next we turn to the convergence of the STC system. Considering Eq. (2.72), it is clear that

$$\frac{1}{n}\sum_{k=1}^{n}|y_2(k)|^2 = o(1) \tag{2.81}$$

Further, by Lemma A.4 in Appendix A, we get

$$\frac{1}{n}\sum_{k=1}^{n}|y(k)-y_r(k)|^2 = \frac{1}{n}\sum_{k=1}^{n}|y_1(k)-y_r(k)+y_2(k)|^2 \to \frac{1}{n}\sum_{k=1}^{n}|y_1(k)-y_r(k)|^2 \tag{2.82}$$

That is

$$\lim_{n\to\infty}\frac{1}{n}\sum_{k=1}^{n}|y(k)-y_r(k)|^2 = \lim_{n\to\infty}\frac{1}{n}\sum_{k=1}^{n}|y_1(k)-y_r(k)|^2 \tag{2.83}$$

That completes the proof of Theorem 2.6. □

2.3.2.2 Parameter Estimates Converge But Are Not Consistent

We have the following result for a general stochastic STC system consisted of arbitrary control strategy and arbitrary estimation algorithm.

Theorem 2.7 *The STC system of stochastic plant with known structure information (i.e., n, m, and d) is stable and convergent, if only the following properties are satisfied:*

(1) The control strategy is well defined such that $\Sigma\{C_0, P_0\}$ constitutes a stable closed-loop system;
(2) $P_m(k)$ is convergent and uniformly controllable with

$$\sum_{k=1}^{n} |y(k) - \phi^T(k-d)\hat{\theta}(k) - \omega(k)|^2 = o(1 + \sum_{k=1}^{n} \|\phi(k-d)\|^2) \tag{2.84}$$

(3) The mapping from estimated parameters into controller parameters is continuous at θ_0.

Proof The VES as shown in Fig. 2.16 can be decomposed into three subsystems, i.e, Figs. 2.19, 2.20, and 2.21, respectively.

According to superposition principle we have

$$y(k) = y_1(k) + y_2(k) + y_3(k), \quad u(k) = u_1(k) + u_2(k) + u_3(k) \tag{2.85}$$

Similar to the proof of Theorem 2.6, define $\widetilde{\phi}(k)$

$$\widetilde{\phi}(k) = \widetilde{\phi}_1(k) + \widetilde{\phi}_2(k) + \widetilde{\phi}_3(k) \tag{2.86}$$

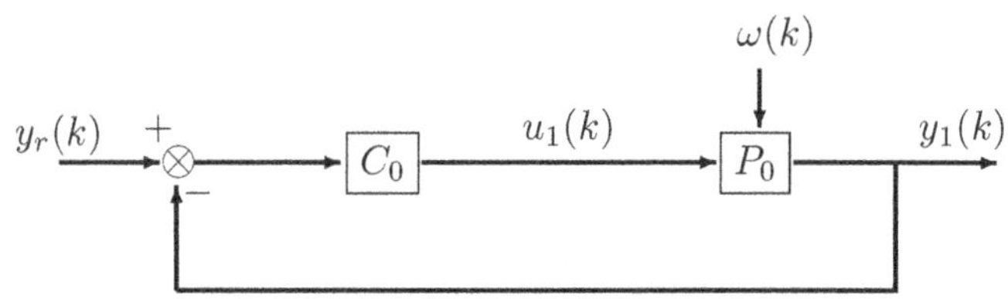

Fig. 2.19 Block diagram for subsystem 1 of stochastic VES II

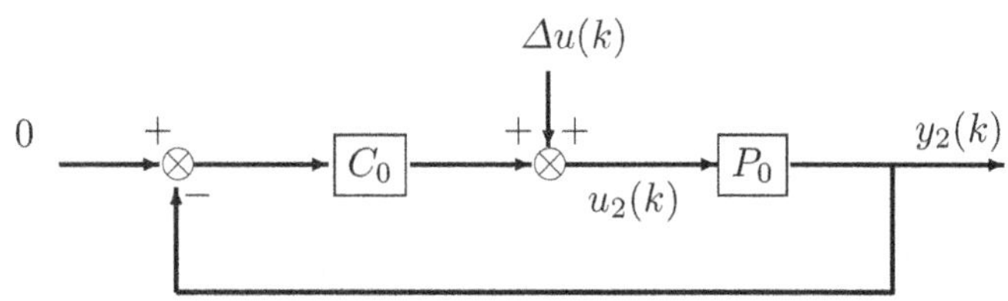

Fig. 2.20 Block diagram for subsystem 2 of stochastic VES II

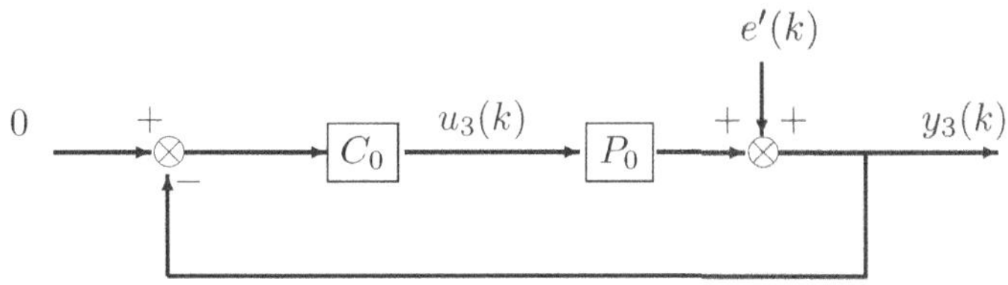

Fig. 2.21 Block diagram for subsystem 3 of stochastic VES II

Then we obtain

$$\|\phi_c(k)\| = O(\|\widetilde{\phi}(k)\|), \quad \|\phi(k-d)\| = O(\|\widetilde{\phi}(k)\|) \tag{2.87}$$

First, by Condition (1), we know that subsystem 1 is a stable stochastic linear time-invariant system, That means

$$\frac{1}{n}\sum_{k=1}^{n} \|\widetilde{\phi}_1(k)\|^2 < \infty, \quad \frac{1}{n}\sum_{k=1}^{n} |y_1(k) - y_r(k)|^2 < \infty \tag{2.88}$$

Second, subsystem 2 (Fig. 2.20) is a stable deterministic system. Condition (2) and Condition (3) guarantee the property of its input signal

$$\Delta u(k) = o(\|\phi_c(k)\|) = o(\|\widetilde{\phi}(k)\|) \tag{2.89}$$

Thus, we have

$$\frac{1}{n}\sum_{k=1}^{n} |y_2(k)|^2 = o\left(\frac{1}{n}\sum_{k=1}^{n} \|\widetilde{\phi}(k)\|^2\right), \quad \frac{1}{n}\sum_{k=1}^{n} |u_2(k)|^2 = o\left(\frac{1}{n}\sum_{k=1}^{n} \|\widetilde{\phi}(k)\|^2\right) \tag{2.90}$$

$$\frac{1}{n}\sum_{k=1}^{n} \|\widetilde{\phi}_2(k)\|^2 = o\left(\frac{1}{n}\sum_{k=1}^{n} \|\widetilde{\phi}(k)\|^2\right) \tag{2.91}$$

Finally, let's consider subsystem 3 (Fig. 2.21). From Condition (2) Eqs. (2.65) and (2.85), we obtain

$$\sum_{k=1}^{n} |e'(k)|^2 = o\left(1 + \sum_{k=1}^{n} \|\phi(k-d)\|^2) = o(1 + \sum_{k=1}^{n} \|\widetilde{\phi}(k)\|^2\right) \tag{2.92}$$

Thus, knowing that subsystem 3 is stable, we have

$$\frac{1}{n}\sum_{k=1}^{n}|y_3(k)|^2 = o\left(\frac{1}{n}\sum_{k=1}^{n}\|\widetilde{\phi}(k)\|^2\right),\quad \frac{1}{n}\sum_{k=1}^{n}|u_3(k)|^2 = o\left(\frac{1}{n}\sum_{k=1}^{n}\|\widetilde{\phi}(k)\|^2\right) \tag{2.93}$$

$$\frac{1}{n}\sum_{k=1}^{n}\|\widetilde{\phi}_3(k)\|^2 = o\left(\frac{1}{n}\sum_{k=1}^{n}\|\widetilde{\phi}(k)\|^2\right) \tag{2.94}$$

Then we obtain the result of Theorem 2.7 following the similar procedures of the proof of Theorem 2.6. □

2.3.2.3 Parameter Estimates May Not Converge

For simplicity of understanding, we first consider minimum variance STC combined with arbitrary parameter estimation algorithm. In this situation we still use VES as shown in Fig. 2.15. Although minimum variance STC has been proved stable and convergent (self-optimality) in [5] with stochastic approximation algorithm, and in [10] with least square algorithm, our analysis is more simple and direct. Second, we will consider a general stochastic STC system, which comprises arbitrary control strategy with arbitrary parameter estimation algorithm. The VES for this situation will take a new form, i.e, stochastic slow switching system.

Theorem 2.8 *The minimum variance STC system of stochastic plant with known structure information (i.e., n, m, and d) is stable and convergent, if only the following properties are satisfied:*

(1) $B(q^{-1})$ *is a Hurwitz polynomial, and* $b_0 \neq 0$*;*
(2) $\hat{b}_0 \neq 0$ *or some modification on* $\hat{b}_0$ *to guarantee the existence of* $u(k)$*;*
(3) Parameter estimates yield

$$\sum_{k=1}^{n}|y(k) - \phi^T(k-d)\hat{\theta}(k) - \omega(k)|^2 = o\left(1 + \sum_{k=1}^{n}\|\phi(k-d)\|^2\right) \tag{2.95}$$

Proof From the processes of proofs of Theorems 2.6 and 2.7, we could see that the key point is the property of $\Delta u(k)$ in VES as shown in Fig. 2.15. For minimum variance control strategy, we have shown in ref. [32] that the convergence of parameter estimation is not necessary to keep the property of $\Delta u(k)$. The deduction procedures are much similar to that in the one-step-ahead STC (see the proof of Theorem 2.3). The special property of $\Delta u(k)$ is owing to the unique characteristic of minimum variance control strategy. In short we have

$$\Delta u(k) = \frac{1}{b_0}[y(k) - \phi^T(k-d)\hat{\theta}(k) - \omega(k)] \tag{2.96}$$

This together with Eq.(2.93) indicates

$$\frac{1}{n}\sum_{k=1}^{n}|\Delta u(k)|^2 = o\left(\frac{1}{n}\sum_{k=1}^{n}\|\phi(k-d)\|^2\right) = o\left(\frac{1}{n}\sum_{k=1}^{n}\|\widetilde{\phi}(k)\|^2\right) \tag{2.97}$$

The remained procedures are similar to the proof of Theorem 2.6. □

Remark 2.5 Condition (3) in Theorem 2.8 is identical to

$$\sum_{k=1}^{n}|\hat{\omega}(k)-\omega(k)|^2 = o\left(1+\sum_{k=1}^{n}\|\phi(k-d)\|^2\right) \tag{2.98}$$

in most literatures.

Theorem 2.9 *The STC system of a stochastic plant with known structure information (i.e., n, m, and d) is stable and convergent, if only the following properties are satisfied:*

(1) Control strategy is well defined to stabilize parameter known plant and tracking the bounded reference signal $y_r(k)$ in mean square sense;
(2) $\|\hat{\theta}(k)\| \le M < \infty$, $\|\hat{\theta}(k)-\hat{\theta}(k-l)\| \to 0$, l is a finite integer.
(3) The parameter estimates satisfy the performance

$$\sum_{k=1}^{n}|y(k)-\phi^T(k-d)\hat{\theta}(k)-\omega(k)|^2 = o\left(1+\sum_{k=1}^{n}\|\phi(k-d)\|^2\right) \tag{2.99}$$

(4) The mapping from model parameters into controller parameters is continuous in $\hat{\theta}(k)$, for $\|\hat{\theta}(k)\| \le M < \infty$.

Remark 2.6 According to Cauchy convergence criterion, condition (2), i. e, $\|\hat{\theta}(k)\| < \infty$, $\|\hat{\theta}(k)-\hat{\theta}(k-l)\| \to 0$, l is a finite integer, does not mean the convergence of $\hat{\theta}(k)$. In fact, $\|\hat{\theta}(k)\| < \infty$, and $\|\hat{\theta}(k)-\hat{\theta}(k-l)\| \to 0$, $l \to \infty$ means the convergence of $\hat{\theta}(k)$.

Proof A stochastic STC system as shown in Fig. 2.14 is obviously equivalent to the system as shown in Fig. 2.22. Further, we obtain an equivalent system as shown in Fig. 2.23, where $\Delta u'(k) = \phi_c^T(k)[\hat{\theta}_c(k) - \hat{\theta}_c(t_k)]$.

Next, the VES IV (Fig. 2.23) can be decomposed into three subsystems, as shown in Figs. 2.24, 2.25, and 2.26, respectively.

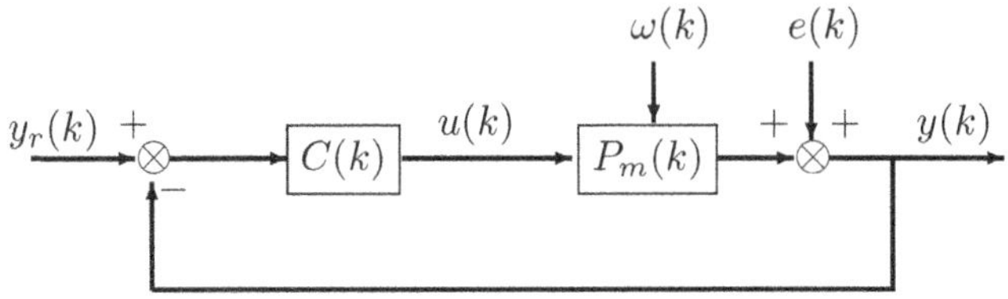

Fig. 2.22 Block diagram for stochastic VES III

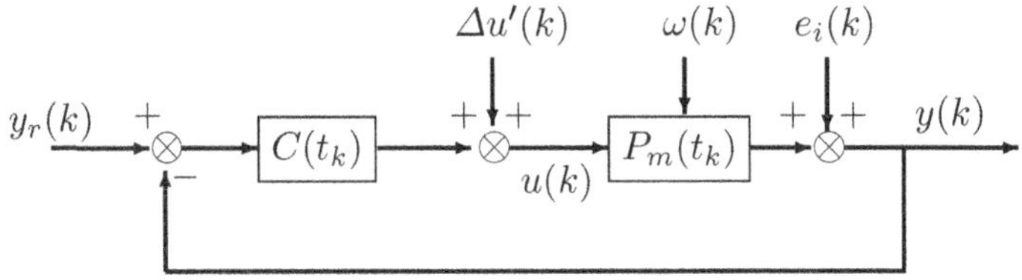

Fig. 2.23 Block diagram for stochastic VES IV

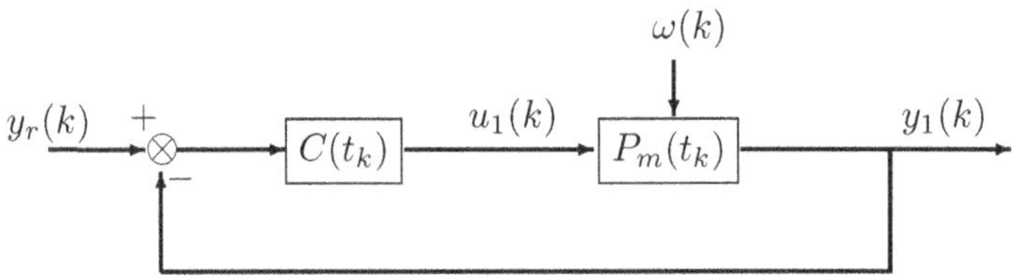

Fig. 2.24 Block diagram for subsystem 1 of stochastic VES IV

Then by superposition principle, we have

$$y(k) = y_1(k) + y_2(k) + y_3(k) \tag{2.100}$$

$$u(k) = u_1(k) + u_2(k) + u_3(k) \tag{2.101}$$

Similar to the proof of Theorem 2.7, we define $\widetilde{\phi}(k)$

$$\widetilde{\phi}(k) = \widetilde{\phi}_1(k) + \widetilde{\phi}_2(k) + \widetilde{\phi}_3(k) \tag{2.102}$$

Thus, it follows that

$$\|\phi_c(k)\| = O(\|\widetilde{\phi}(k)\|), \quad \|\phi(k-d)\| = O(\|\widetilde{\phi}(k)\|) \tag{2.103}$$

The choice of sequence t_k is the same as in the proof of Theorem 2.4 in Sect. 2.2.2.3, which together with Condition (2) guarantees the VES IV (Fig. 2.23) is a "slow switching" system with large enough dwell time. Then subsystem 1 (Fig. 2.24) is a

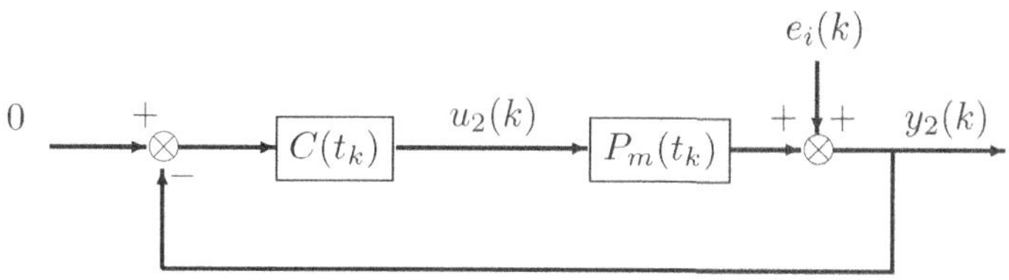

Fig. 2.25 Block diagram for subsystem 2 of stochastic VES IV

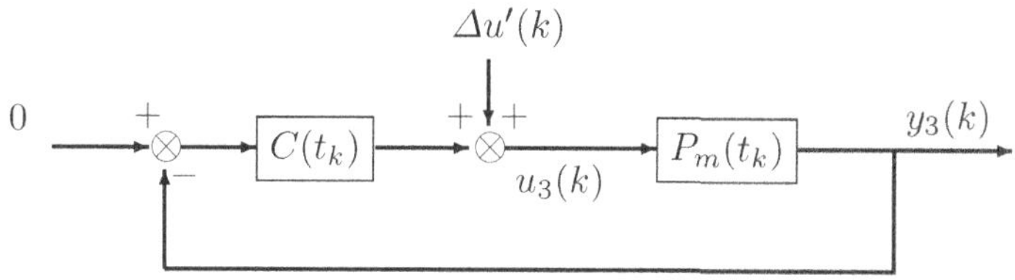

Fig. 2.26 Block diagram for subsystem 3 of stochastic VES IV

stochastic slow switching system, and the other two subsystems (Figs. 2.25 and 2.26) are deterministic slow switching systems.

Further, according to Condition (1) we know that for each given t_k, $C(t_k)$ and $P_m(t_k)$ constitute a stable closed-loop system. Then, based on the existing results of "slow switching" system [41–44], we know that all the three subsystems are stable. That means in subsystem 1 (Fig. 2.24)

$$\frac{1}{n}\sum_{k=1}^{n}\|\widetilde{\phi}_1(k)\|^2 < \infty \tag{2.104}$$

In subsystem 2 (Fig. 2.25), its input signal is given by

$$\begin{aligned} e_i(k) &= y(k) - \phi^T(k-d)\hat{\theta}_{t_k} - \omega(k) \\ &= y(k) - \phi^T(k-d)\hat{\theta}(k) - \omega(k) + \phi^T(k-d)[\hat{\theta}_k - \hat{\theta}_{t_k}] \end{aligned} \tag{2.105}$$

Taking account of Condition (2) and Condition (3), we get

$$\sum_{k=1}^{n}|e_i(k)|^2 = o\left(1 + \sum_{k=1}^{n}\|\phi(k-d)\|^2\right) = o\left(1 + \sum_{k=1}^{n}\|\widetilde{\phi}(k)\|^2\right) \tag{2.106}$$

Then it follows that

$$\frac{1}{n}\sum_{k=1}^{n}|y_2(k)|^2 = o\left(\frac{1}{n}\sum_{k=1}^{n}\|\widetilde{\phi}(k)\|^2\right), \quad \frac{1}{n}\sum_{k=1}^{n}|u_2(k)|^2 = o\left(\frac{1}{n}\sum_{k=1}^{n}\|\widetilde{\phi}(k)\|^2\right) \tag{2.107}$$

$$\frac{1}{n}\sum_{k=1}^{n}\|\widetilde{\phi}_2(k)\|^2 = o\left(\frac{1}{n}\sum_{k=1}^{n}\|\widetilde{\phi}(k)\|^2\right) \tag{2.108}$$

In subsystem 3 (Fig. 2.26), by Conditions (2) and (4), its input signal has the following property

$$\Delta u'(k) = o(\|\phi_c(k)\|) = o(\|\widetilde{\phi}(k)\|) \tag{2.109}$$

Then we have

$$\frac{1}{n}\sum_{k=1}^{n}|y_3(k)|^2 = o\left(\frac{1}{n}\sum_{k=1}^{n}\|\widetilde{\phi}(k)\|^2\right), \quad \frac{1}{n}\sum_{k=1}^{n}|u_3(k)|^2 = o\left(\frac{1}{n}\sum_{k=1}^{n}\|\widetilde{\phi}(k)\|^2\right) \tag{2.110}$$

$$\frac{1}{n}\sum_{k=1}^{n}\|\widetilde{\phi}_3(k)\|^2 = o\left(\frac{1}{n}\sum_{k=1}^{n}\|\widetilde{\phi}(k)\|^2\right) \tag{2.111}$$

Following the similar procedures of the stability proof of Theorem 2.6, i.e., by way of contradiction, we come to the conclusion that the STC system is stable.

Next we proceed to prove the convergence of the STC system. First we examine the tracking performance of subsystem 1. Denote the tracking error of each "frozen" system $\sum\{C(\bar{t}_k), P_m(\bar{t}_k), \omega(k)\}$ by

$$e_{\bar{t}_k}(k) = y_{\bar{t}_k}(k) - y_r(k)$$

where $\bar{t}_k$ represents the "frozen" t_k.

Similarly, the tracking error of $\sum\{C(t_k), P_m(t_k), \omega(k)\}$, i.e., subsystem 1 (Fig. 2.24), is denoted by

$$e_{t_k}(k) = y_{t_k}(k) - y_r(k) = y_1(k) - y_r(k)$$

Then we know that for each "frozen" system $\sum\{C(\bar{t}_k), P_m(\bar{t}_k), \omega(k)\}$, there exists a stable dynamic system

$$e_{\bar{t}_k}(k+1) = f_{\bar{t}_k}(e_{\bar{t}_k}(k), \omega(k), k)$$

Subsequently, by MLF methodology, we know that for $\sum\{C(t_k), P_m(t_k), \omega(k)\}$, there exists a stable "slow switching" dynamic system.

$$e_{t_k}(k+1) = f_{t_k}(e_{t_k}(k), \omega(k), k)$$

According to Condition (1), for each $\bar{t}_k$, we have

$$\lim_{n\to\infty} sup\frac{1}{n}\sum_{k=1}^{n}|e_{\bar{t}_k}(k)|^2 = R$$

It is easy to see that $e_{t_k}(k)$ takes values in the set of $e_{\bar{t}_k}(k)$, then by squeezing, it follows that

$$\lim_{n\to\infty} sup\frac{1}{n}\sum_{k=1}^{n}|e_{t_k}(k)|^2 = R$$

That implies the tracking performance of the "slow switching" subsystem 1, i.e.

$$\lim_{n\to\infty} sup\frac{1}{n}\sum_{k=1}^{n}|y_1(k) - y_r(k)|^2 = R$$

Next, following the similar procedures of the convergence proof of Theorem 2.6, we see that the self-tung control system is convergent.

Remark 2.7 Instead of requiring the convergence of parameter estimates, Theorem 2.9 actually only requires asymptotic property of parameter estimates. Compared with Theorem 2.8, Condition (2) of Theorem 2.9 is a little strict (i.e. $\|\hat{\theta}(k)\| < \infty$). It has been known that Stochastic Gradient (SG) algorithm and Weighted Extended Least Squares (WELS) both have the properties required by Theorem 3.4, i.e., Conditions (2) and (3), for more details, see reference [22].

We now extend the above result, i.e., Theorem 2.9 to the stochastic plant with colored noise.

Consider

$$A(q^{-1})y(k) = q^{-d}B(q^{-1})u(k) + D(q^{-1})\omega(k) \tag{2.112}$$

where

$$\begin{aligned} A(q^{-1}) &= 1 + a_1q^{-1} + \cdots + a_nq^{-n} \\ B(q^{-1}) &= b_0 + b_1q^{-1} + \cdots + b_mq^{-m} \\ D(q^{-1}) &= 1 + d_1q^{-1} + \cdots + d_sq^{-s} \end{aligned}$$

Corollary 2.1 *The STC system of stochastic plant with known structure information (i.e., n, m, and d) is stable and convergent, if only the following properties are satisfied:*

(1) Control strategy is well defined to stabilize parameter known plant and tracking the bounded reference signal $y_r(k)$ in mean square sense;
(2) $\|\hat{\theta}(k)\| \leq M < \infty$, $\|\hat{\theta}(k) - \hat{\theta}(k-l)\| \to 0$, l is a finite integer.
(3) The parameter estimates have the property

$$\sum_{k=1}^{n}|y(k)-\phi^T(k-d)\hat{\theta}(k)-\omega(k)|^2=o\left(1+\sum_{k=1}^{n}\|\phi(k-d)\|^2\right) \quad (2.113)$$

(4) The mapping from model parameters into controller parameters is continuous in $\hat{\theta}(k)$, for $\|\hat{\theta}(k)\|\le M<\infty$.

Proof Denote

$$\theta^T=[a_1,\ldots,a_n,b_0,b_1,\ldots,b_m,d_1,\ldots,d_s]$$

To estimate the parameters of $D(q^{-1})$, we need to estimate

$$\hat{\omega}(k)=y(k)-\phi^T(k-d)\hat{\theta}(k) \quad (2.114)$$

$$\begin{aligned}\phi^T(k-d)=&[y(k-1),\ldots,y(k-n),u(k-d),\ldots,u(k-d-m),\\&\hat{\omega}(k-1),\ldots,\hat{\omega}(k-s)]\end{aligned} \quad (2.115)$$

We will still use Fig. 2.22 (further, Fig. 2.23) as the VESs for this situation, but the definition of $e(k)$ is different from before, i.e.

$$e(k)=y(k)-[\phi_0^T(k-d)\hat{\theta}(k)+\omega(k)]$$

where

$$\phi_0^T(k-d)=[y(k-1),\ldots,y(k-n),u(k-d),\ldots,u(k-d-m),\omega(k-1),\ldots,\omega(k-s)]$$

It is not difficult to imagine that if only the properties of $e_i(k)$ and $\Delta u'(k)$ remain the same as in the proof of Theorem 2.9, then the Corollary holds.

We have

$$\begin{aligned}e(k)&=y(k)-[\phi^T(k-d)\hat{\theta}(k)+\omega(k)]+\phi^T(k-d)\hat{\theta}(k)-\phi_0^T(k-d)\hat{\theta}(k)\\&=\hat{\omega}(k)-\omega(k)+e'(k)\end{aligned} \quad (2.116)$$

where

$$\begin{aligned}e'(k)&=\phi^T(k-d)\hat{\theta}(k)-\phi_0^T(k-d)\hat{\theta}(k)\\&=\hat{d}_1[\hat{\omega}(k-1)-\omega(k-1)]+\cdots+\hat{d}_s[\hat{\omega}(k-s)-\omega(k-s)]\end{aligned} \quad (2.117)$$

From Condition (2), it is obvious that

$$\sum_{k=1}^{n}|e'(k)|^2=O\left(\sum_{k=1}^{n}|\hat{\omega}(k)-\omega(k)|^2\right)=o\left(1+\sum_{k=1}^{n}\|\phi(k-d)\|^2\right)$$

This together with (2.112) gives

$$\sum_{k=1}^{n}|e(k)|^2 = O\left(\sum_{k=1}^{n}|\hat{\omega}(k)-\omega(k)|^2+\sum_{k=1}^{n}|e'(k)|^2\right) = o\left(1+\sum_{k=1}^{n}\|\phi(k-d)\|^2\right)$$

Further, by Condition (2) we know the property of $e_i(k) = e(k) + \phi^T(k-d)[\hat{\theta}_k - \hat{\theta}_{t_k}]$ in VES Fig. 2.23 remains the same as in the proof of Theorem 2.9 although colored noise is considered.

As for $\Delta u'(k)$, by Conditions (2) and (4), it is easy to see that $\Delta u'(k) = o(\|\phi_c(k)\|)$.

Therefore, the Corollary holds.

Remark 2.8 The other Theorems, i.e., Theorems 2.6, 2.7 and 2.8 could also be extended to plant (2.110) similarly. Details are omitted.

Next we consider an LTV system, i.e., parameter jumping system

$$A(\gamma, q^{-1})y(k) = q^{-d}B(\gamma, q^{-1})u(k) + D(\gamma, q^{-1})\omega(k) \tag{2.118}$$

where

$$\begin{aligned}A(\gamma, q^{-1}) &= 1 + a_1(\gamma)q^{-1} + \cdots + a_n(\gamma)q^{-n}\\ B(\gamma, q^{-1}) &= b_0(\gamma) + b_1(\gamma)q^{-1} + \cdots + b_m(\gamma)q^{-m}\\ D(\gamma, q^{-1}) &= 1 + d_1(\gamma)q^{-1} + \cdots + d_s(\gamma)q^{-s}\end{aligned}$$

Denote

$$\theta^T(\gamma) = [a_1(\gamma), \ldots, a_n(\gamma), b_0(\gamma), b_1(\gamma), \ldots, b_m(\gamma), d_1(\gamma), \ldots, d_s(\gamma)]$$

We need the following assumptions:

(A1) $\theta(\gamma)$ belongs to a compact subset of a finite-dimensional linear vector space, i.e. $\theta(\gamma) \in \Theta \subset \mathbb{R}^{n+m+s+1}$.

(A2) $a_i(\gamma), i = 1, \ldots, n,\ b_j(\gamma), j = 0, 1, \ldots, m,$ and $d_l(\gamma), l = 1, \ldots, s$ are all piecewise constant; the interval between consecutive parameter jumps is large enough.

(A3) The estimation algorithm can detect the parameters jump, say, by measuring the jump of estimation error $e(k) = y(k) - \phi^T(k-d)\hat{\theta}(k)$, and then the estimator can be reinitialized.

Then we have the following result.

Theorem 2.10 *Assume (A1) (A2) and (A3) hold. If the STC system of stochastic plant (2.116) with known structure information, i.e., n, m, s and d, has the following properties*

(1) Control strategy is well defined to stabilize parameter known plant and tracking the bounded reference signal $y_r(k)$;
(2) For each "frozen" γ or $\theta(\gamma)$, $\|\hat{\theta}(k)\| \leq M < \infty$, and $\|\hat{\theta}(k) - \hat{\theta}(k-l)\| \to 0$, l is a finite integer.
(3) For each "frozen" γ or $\theta(\gamma)$, the estimation error satisfies:

$$\sum_{k=1}^{n} |\hat{\omega}(k) - \omega(k)|^2 = o\left(1 + \sum_{k=1}^{n} \|\phi(k-d)\|^2\right)$$

(4) The mapping from model parameters into controller parameters is continuous in $\hat{\theta}(k)$, for $\|\hat{\theta}(k)\| \leq M < \infty$.
Then the STC system is stable.

Proof According to Corollary 2.1, Conditions (1)–(4) guarantee that for each "frozen" γ or $\theta(\gamma)$ in (2.116), there exists a stable stochastic STC system denoted by $\sum\{C(k), P_{\theta(\gamma)}\}$. Further, Assumptions (A2) and (A3) guarantee the "slow switching" between these stable control systems with large enough dwell time. Then, by MLF methodology plus Assumption (A1), the overall STC system $\sum\{C(k), P_{\Theta}\}$ is stable.

That completes the proof.

2.4 Simulation Verification of VES

In this section, we verify the equivalence of VES (as shown in Fig. 2.15) before and after decomposition.

Consider the following plant

$$\begin{aligned} &y(k) - 1.6065y(k-1) + 0.6065y(k-2) \\ &= 0.1065u(k-10) + 0.0902u(k-4) + \omega(k) + 0.5\omega(k-1) \end{aligned} \tag{2.119}$$

where $\omega(k)$ is white noise, with variance $\sigma^2 = 0.01$. In the STC system, least squares estimation algorithm is adopted, with $P(0) = 10^6 I$, $\hat{\theta}(0) = \mathbf{0}$, the control strategy is pole-placement. And the desired closed-loop poles are selected as follows

$$A_m(q^{-1}) = 1 - 1.3205q^{-1} + 0.4966q^{-2}$$

The simulation results are given in Figs. 2.27 and 2.28, respectively. Figure 2.27 includes $y_1(k)$, $y_2(k)$ and $y_1(k) + y_2(k)$ versus $y(k)$, owing to the red color (plotted former) of $y_1(k) + y_2(k)$ and the blue color (plotted later) of $y(k)$ are totally the same,

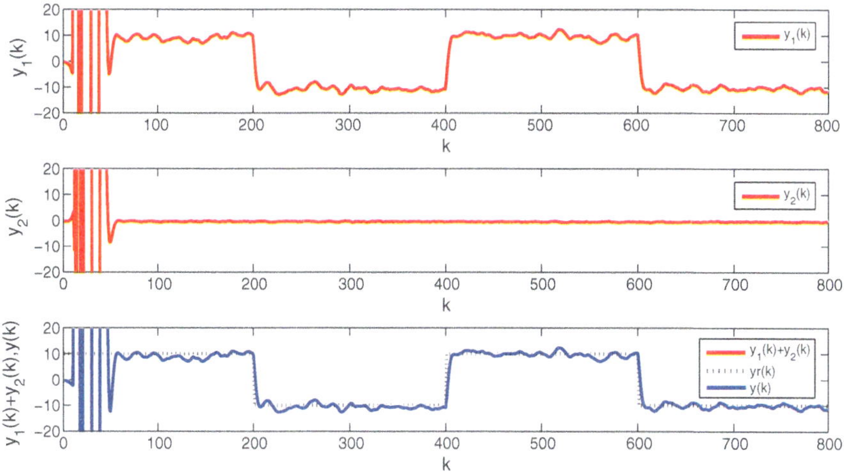

Fig. 2.27 Equivalence of output signals of VES before and after decomposition

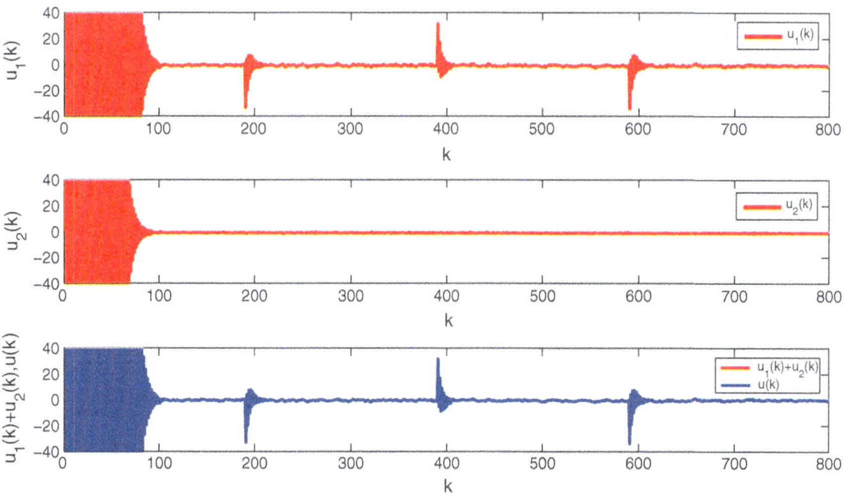

Fig. 2.28 Equivalence of control signals of VES before and after decomposition

so that we can see only the blue line, that verifies the fact $y(k) = y_1(k) + y_2(k)$; Fig. 2.28 gives the comparison of $u_1(k) + u_2(k)$ and $u(k)$. Similarly to Figs. 2.27, 2.28 verifies that $u(k) = u_1(k) + u_2(k)$.

2.5 Guidelines to Designing an STC System

Through the analysis results presented in the former sections, we may obtain some new perspectives on STC systems. First, the stability and convergence of an STC system depends on the accuracy of the estimated model as a whole, i.e., $e(k)$ or $\sum_{k=1}^{n} \|e(k) - \omega(k)\|^2$, rather than the accuracy of every single estimated parameter. Second, the consistency or the convergence of parameter estimates is not necessary, i.e., persistent excitation condition is not necessary for the stability and convergence of a general STC system provided that the structure information of plant is known. Third, the adjustment of controller parameters makes disturbances to the system, they are $\Delta u(k)/\Delta u'(k)$ and $e(k)/e_i(k)$. That is the cost of adaptive or learning. How to make an STC system robust? i.e., how to attenuate the disturbances? We propose the following guidelines:

a. The parameter estimator and control strategy are both changeable, i.e., we may use different control strategies and estimation algorithms at different stages of adaptive control. For example, at the beginning we may use fast algorithm like least squares, and at the steady stage we may use slow algorithm like gradient algorithm to guarantee the stability and convergence of the closed-loop system.
b. Robust estimator to attenuate $e(k)$ plus robust controller to attenuate $\Delta u(k)$.
c. Multi-model adaptive control methods: When $e(k)$ is large we should better not use the estimated model, instead, we may use prior knowledge to formulate the control law. That is to say using Multi-model adaptive control, or even simpler, PID or fuzzy controller in the beginning, and switch to the STC after $e(k)$ is small enough.
d. Intelligent adaptive control method: to learn and predict $e(k)$ and then take it into account in control law, actually this is similar to the thinking of predictive control.

2.6 Conclusion

With the help of VES concept and methodology, we present a unified understanding and judgment of the stability and convergence of STC systems. The necessary conditions for the stability and convergence have been relaxed.

Also we may answer the question which was put forward in the beginning of this chapter. Why minimum variance STC can guarantee the stability and convergence without the convergence of parameter estimates, but other self-tuning control schemes, like pole placement or LQG, cannot? It is not caused by the differences of control strategies or the characteristic of plant to be controlled. It is caused by the lack of appropriate analysis tool. Actually there are no big differences between various types of STC schemes in the sense of stability and convergence.

References

1. K.J. Åström, B. Wittenmark, On self-tuning regulators. Automatica **9**, 185–199 (1973)
2. D.W. Clarke, P.J. Gawthrop, Self-tuning controller. Proc. IEE **122**(9), 929–934 (1975)
3. P.E. Wellstead, J.M. Edmunds, D. Prager, P. Zanker, Self-tuning pole/zero assignment regulators. Int. J. Control **30**(1), 1–26 (1979)
4. P.J. Gawthrop, On the stability and convergence of a self-tuning controller. Int. J. Control **31**(5), 973–998 (1980)
5. G.C. Goodwin, P.J. Ramadge, P.E. Caines, Discrete time stochastic adaptive control. SIAM J. Control Optim. **19**, 829–853 (1981)
6. D.W. Clarke, Self-tuning control of non-minimum-phase systems. Automatica **20**(5), 501–517 (1984)
7. F. Giri, M. M'Saad, L. Dugard, J.M. Dion, Robust pole placement indirect adaptive control. Int. J. Adapt. Control Signal Process. **2**, 33–47 (1988)
8. G. Kreisselmeier, An indirect adaptive controller with a self-excitation capability. IEEE Trans. Automatic Control **34**, 524–528 (1989)
9. P.R. Kumar, Convergence of adaptive control schemes using least squares parameter estimates. IEEE Trans. Automatic Control **35**(4), 416–424 (1990)
10. L. Guo, H.F. Chen, The Astrom-Wittenmark self-tuning regulator revisited and ELS-based adaptive trackers. IEEE Trans. Automatic Control **36**, 802–812 (1991)
11. H.F. Chen, J.F. Zhang, Adaptive stabilization of unstable and non-minimum phase stochastic systems. Syst. Control Lett. **20**, 27–38 (1993)
12. R. Lozano, X.H. Zhao, Adaptive pole placement without excitation probing signals. IEEE Trans. Automatic Control **3**(9), 47–58 (1994)
13. C.J. Harris, S.A. Billings, *Self-Tuning and Adaptive Control: Theory and Applications* (Peter Peregrinus Ltd., 1981)
14. G.C. Goodwin, K.S. Sin, *Adaptive Filtering, Prediction and Control* (Prentice Hall, 1984)
15. S. Sastry, M. Bodson, *Adaptive Control: Stability, Convergence and Robustness* (Prentice Hall, Englewood Cliffs, New Jersey, 1989)
16. K.J. Åström, B. Wittenmark, *Adaptive Control* (Addison-Wesley, 1995)
17. P.A. Ioannou, J. Sun, *Robust Adaptive Control* (Prentice-Hall: Englewood Cliffs, USA, New Jersey, 1996)
18. G. Tao, *Adaptive Control Design and Analysis* (Wiley, New York, 2003)
19. G.C. Goodwin, A. Feuer, Adaptive control: where to now? adaptive systems for signal processing, communications, and control symposium, in *Proceedings IEEE* (2000), pp. 165–170
20. S. Fekri, M. Athans, A. Pascoal, Issues, progress and new results in robust adaptive control. Int. J. Adapt. Control Signal Process **20**(10), 519–579 (2006)
21. J.H. Van Schuppen, Tuning of gaussian stochastic control systems. IEEE Trans. Automatic Control **39**(11), 2178–2190 (1994)
22. K. Nassiri-Toussi, W. Ren, A unified analysis of stochastic adaptive control: asymptotic self-tuning, in *Proceedings of the 34th IEEE Conference on Decision and Control* (1995) pp. 2932–2937
23. A.S. Morse, Towards a unified theory of parameter adaptive control-part II: certainty equivalence and implicit tuning. IEEE Trans. Automatic Control **37**(1), 15–29 (1992)
24. L. Guo, Self-convergence of weighted least—squares with applications to stochastic adaptive control. IEEE Trans. Automatic Control **41**(1), 79–89 (1996)
25. K. Nassiri-Toussi, W. Ren, Indirect adaptive pole-placement control of MIMO stochastic systems: self-tuning results. IEEE Trans. Automatic Control **42**(1), 38–52 (1997)
26. T. Katayama, T. McKelvey, A. Sano, C.G. Cassandras, M.C. Campi, Trends in systems and signals: status report prepared by the IFAC coordinating committee on systems and signals. Ann. Rev. Control **30**, 5–17 (2006)
27. W. Zhang, Theoretical research and application of robust adaptive control, Ph.D. Thesis (Tsinghua University, 1993)

28. W. Zhang, J.Y. Choi, On the stability and convergence of self-tuning control systems (I)—deterministic plant, in *International Conference on Control, Automation and Systems* (Seoul, Korea, 2007)
29. W. Zhang, J.Y. Choi, On the stability and convergence of self-tuning control systems (II)—stochastic plant, in *International Conference on Control, Automation and Systems* (Seoul, Korea, 2007)
30. W. Zhang, X. Li, J.Y. Choi, A unified analysis of switching multiple model adaptive control—Virtual equivalent system approach, in *17th IFAC Word Congress* (Seoul, Korea, 2008)
31. W. Zhang, X. Li, T. Chu, Virtual equivalent system method for analysis of stochastic self-tuning control, in *The 6th World Congress on Intelligent Control and Automation* (Dalian, China, 2006)
32. W. Zhang, On the stability and convergence of self-tuning control-virtual equivalent system approach. Int. J. Control **83**(5), 879–896 (2010)
33. B. Egardt, Unification of some discrete time adaptive control schemes. IEEE Trans. Automatic Control **25**, 693–697 (1980)
34. P.J. Gawthrop, Some interpretations of the self-tuning controller. Proc. IEEE **124**, 889–894 (1977)
35. L. Ljung, I.D. Landau, Model reference adaptive systems and self-tuning regulators—some connections, in *Proceedings 7th IFAC Congress*, vol. 3 (1978), pp. 1973–1980
36. K.S. Narendra, L.S. Valavani, Direct and indirect adaptive control. Automatica **15**, 663–664 (1979)
37. R. Shorten et al., Stability criteria for switched and hybrid systems. SIAM Rev. **49**, 545–592 (2007)
38. D. Liberzon, A.S. Morse, Basic problems in stability and design of switched systems. IEEE Control Syst. Mag. **19**, 59–70 (1999)
39. J.H. Frommer, S.R. Kulkarni, P.J. Ramadge, Controller switching based on output prediction errors. IEEE Trans. Automatic Control **43**, 596–607 (1998)
40. D. Chatterjee, D. Liberzon, Stability analysis of deterministic and stochastic switched systems via a comparison principle and multiple Lyapunov functions. SIAM J. Control Optim. **45**(1), 174–206 (2006)
41. D. Chatterjee, D. Liberzon, On stability of stochastic switched systems, in *Proceedings of 43rd Conference on Decision and Control*, vol. 4 (2004), pp. 4125–4127
42. M.C. Campi, J.P. Hespanha, M. Prandini, Cautious hierarchical switching control of stochastic linear systems. Int. J. Adapt. Control Signal Process. **18**, 319–333 (2004)
43. M. Prandini, Switching control of stochastic linear systems: Stability and performance results, in *Proceedings 6th Congress of SIMAI* (Chia Laguna, Cagliari, Italy, 2002)
44. M. Prandini, M.C. Campi, Logic-based switching for the stabilization of stochastic systems in presence of unmodeled dynamics, in *Proceedings 40th CDC* (Orlando, USA, 2001)

Chapter 3
Further Results on Stability and Convergence of Self-Tuning Control Systems

Abstract This chapter extends the results of STC systems in Chap. 2 to some more general cases, i.e., the plant may be linear or nonlinear, time-invariant or time varying, or even with unmodelled dynamics. Actually we didn't put any limitations on the plant to be controlled; but it is indirectly restricted in some conditions on the performances of parameter estimates. Roughly speaking, the plant needs to be approximated by linear or piecewise linear models. The necessary conditions for global stability and convergence are relaxed, i.e., the convergence of parameter estimates and the structure information of the plant are both removed. It is therefore suggested that external excitation signal is not necessary for STC system. In other words, feedback information of STC system itself is sufficient to achieve the control objective.

3.1 Introduction

The idea of STC originated from Kalman [1] It was first made applicable by Astrom and Wittenmark [2]. Later on, many books have been published on the subject, see, e.g. [3–11]; Simultaneously, thousands research papers have also been published to deal with different problems of self-tuning control, see, e.g. [12–25], just to name a few.

References [26–29] represents an effort towards a general theory of STC, i.e., a unified understanding and judgment of the stability and convergence of various STC systems, which may consist of arbitrary control strategy, arbitrary parameter estimation algorithm, and a deterministic/stochastic linear time-invariant (LTI) plant with known structure information. In [29], the necessary conditions required for global stability and convergence of a general STC system are relaxed, i.e., the convergence of parameter estimates is removed for both deterministic and stochastic STC schemes. The key point for the progress is mainly due to the "virtual equivalent system (VES)" concept and methodology. With the help of VES, the original nonlinear dominant (nonlinear in structure) problem is converted to a linear dominant (linear in structure) problem. The results developed in [29] show that STC systems are stable and convergent for abundance of control strategies and parameter estimation algorithms, which will provide great flexibility in the applications of self-tuning control.

W. Zhang and Q. Li, *Virtual Equivalent System Approach for Stability Analysis of Model-based Control Systems*, https://doi.org/10.1007/978-981-15-5538-1_3

It should be pointed out that in Chap. 2, we analyzed the stability and convergence of STC systems under condition of a prior knowledge of structure information of the plant to be controlled. But it is well-known that there are always some unmodelled dynamics of the plant to be controlled in practical engineering. So, it is of more significance to develop stability and convergence results for more general cases, such as, structure information of plant is unknown, or even with nonlinearity and time-varying characteristic.

In this chapter, we will consider STC systems for the above-mentioned cases. Actually we didn't put any limitations on the plant to be controlled; but it is indirectly restricted in some conditions on the performances of parameter estimation.

Some lemmas for Chap. 2 will be reused without further proof. In fact, those lemmas remain unchanged even if the structure information of plant is unknown. In other words, those lemmas are independent of system structure information.

3.2 Deterministic STC

3.2.1 Construction of VES

Consider a single input single output (SISO) deterministic plant P with unknown structure information and unknown parameters; it may be linear or non-linear, time-invariant or time varying, or even with unmodelled dynamics. Of course we need to decide the structure information of the plant when we design the STC system. But as analysis methodology, we didn't put any limitations on the plant; it is indirectly restricted in some certain conditions on the performances of parameter estimation.

A general STC system for the above-mentioned plant P is shown in the block diagram of Fig. 3.1 (the estimator is omitted for simplicity).

In Fig. 3.1, $y(k)$, $u(k)$, and $y_r(k)$ are the measurable plant output, control input, and given bounded reference signal, respectively, $y(k) = 0$, $u(k) = 0$, $y_r(k) = 0$ for $k < 0$. $C(k)$ denotes the self-tuning controller at time instant k, it may be designed by any existing principle based on estimated linear model $P_m(k)$, such as by pole placement. The parameterizations of $C(k)$ and $P_m(k)$ are $\theta_c(k)$ and $\hat{\theta}(k)$, respectively. To be specific, $C(k)$ is determined by a mapping between model and controller parameter sets, i.e.

$$f : \mathbb{P} \to \mathbb{C} \tag{3.1}$$

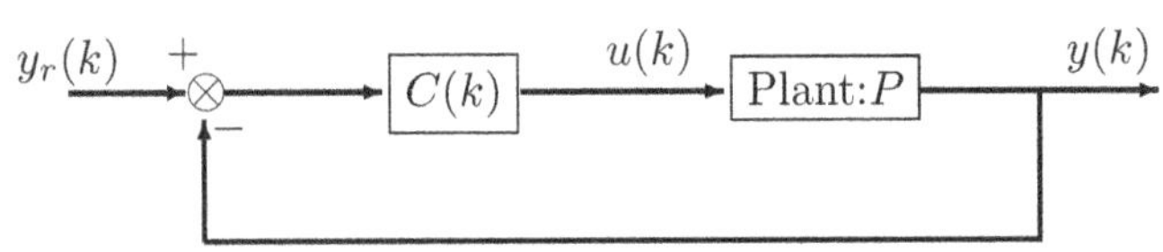

Fig. 3.1 Block diagram for deterministic self-tuning control system

where $\mathbb{P}$ is the model parameter set; $\mathbb{C}$ is the corresponding controller parameter set. Then we have

$$C(k) = f(P_m(k)) \tag{3.2}$$

or

$$\theta_c(k) = f(\hat{\theta}(k)) \tag{3.3}$$

$P_m(k)$ represents the estimated model of plant P at time instant k:

$$\hat{A}(k, q^{-1})y(k) = q^{-d}\hat{B}(k, q^{-1})u(k) \tag{3.4}$$

where $\hat{A}(k, q^{-1})$ and $\hat{B}(k, q^{-1})$ are polynomials in backward-shift operator q^{-1}.

$$\hat{A}(k, q^{-1}) = 1 + \hat{a}_1 q^{-1} + \cdots + \hat{a}_n q^{-n}$$
$$\hat{B}(k, q^{-1}) = \hat{b}_0 + \hat{b}_1 q^{-1} + \cdots + \hat{b}_m q^{-m}$$

For simplicity, we denote

$$P_m(k) = \sum[\hat{A}(k, q^{-1}), \hat{B}(k, q^{-1}), d, n, m] \tag{3.5}$$

The model output $y_m(k)$ can be expressed by

$$y_m(k) = \phi^T(k-d)\hat{\theta}(k) \tag{3.6}$$

where

$$\phi^T(k-d) = [y(k-1), \ldots, y(k-n), u(k-d), \ldots, u(k-d-m)] \tag{3.7}$$

is the data vector (regressor).

$$\hat{\theta}^T(k) = [-\hat{a}_1, \ldots, -\hat{a}_n, \hat{b}_0, \hat{b}_1, \ldots, \hat{b}_m] \tag{3.8}$$

is the the vector of model parameters. Thus, we have

$$y(k) = \phi^T(k-d)\hat{\theta}(k) + e(k) = y_m(k) + e(k) \tag{3.9}$$

where $e(k)$ is the a posteriori estimation (prediction) error

$$e(k) = y(k) - \phi^T(k-d)\hat{\theta}(k) \tag{3.10}$$

The output of the controller is

$$u(k) = \phi_c^T(k)\theta_c(k) \tag{3.11}$$

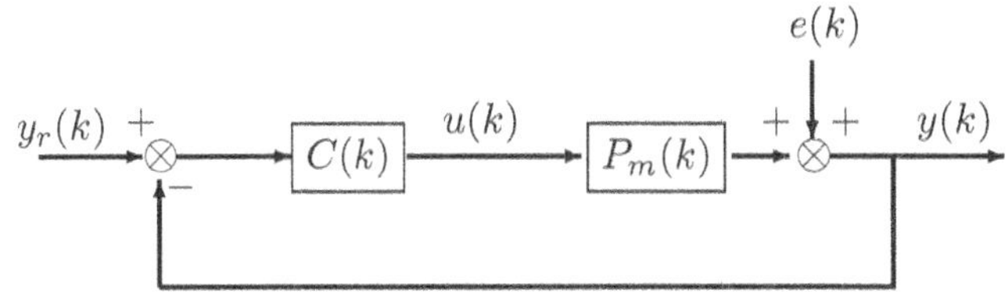

Fig. 3.2 Block diagram for deterministic VES I

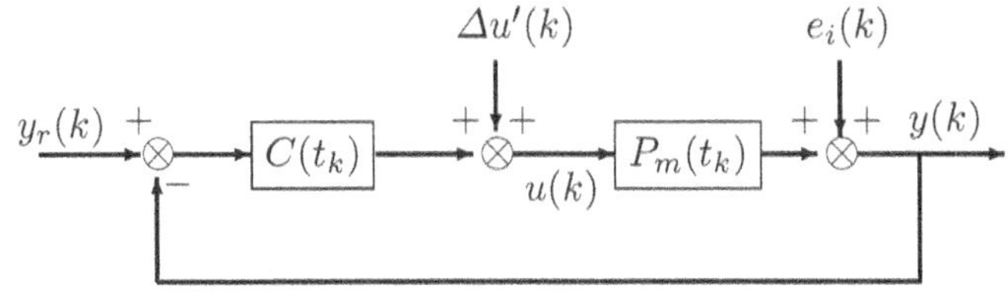

Fig. 3.3 Block diagram for deterministic VES II

where

$$\phi_c^T(k) = [y(k), y(k-1), \ldots, y(k-s_1), u(k-1), \ldots, u(k-s_2), y_r(k), \ldots, y_r(k-s_3)] \tag{3.12}$$

is the vector of controller data, and $\theta_c(k)$ is the vector of controller parameters. The elements of $\phi_c(k)$ and the limited integers s_1, s_2, s_3 depend on specific control strategy.

An STC system (as shown in Fig. 3.1) is obviously equivalent to the system as shown in Fig. 3.2 in the input-output sense. Further, under some mild conditions, we proceed to construct another VES—a slow switching system, which will be easier to deal with. Suppose $\|\hat{\theta}(k)\| \le M < \infty$, $\|\hat{\theta}(k) - \hat{\theta}(k-l)\| \to 0$, l is a finite integer; the mapping $f(.)$ is uniformly continuous on the estimated parameter set. Then we know that $\theta_c(k)/C(k)$ and $\hat{\theta}(k)/P_m(k)$ constitute a slow time varying system. Consequently, we may pick out a subsequence t_k from sequence k such that $C(t_k)$ and $P_m(t_k)$ constitute a slow switching system. Such thinking leads to another VES as shown in Fig. 3.3.

In Fig. 3.3 we have

$$\Delta u'(k) = \phi_c^T(k)[\theta_c(k) - \theta_c(t_k)] \tag{3.13}$$

$$e_i(k) = y(k) - \phi^T(k-d)\hat{\theta}(t_k) \tag{3.14}$$

Figure 3.4 illustrates the choice of subsequence t_k.

To be more precise, we introduce the following definition of t_k

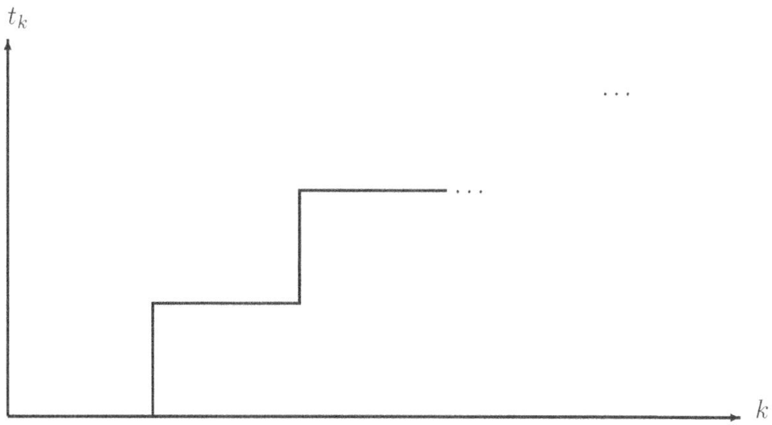

Fig. 3.4 Demonstration of the choice of t_k

$$t_{k+i} = \begin{cases} k+i & \text{if } i = \tau_D \\ t_k & \text{if } i < \tau_D \end{cases} \tag{3.15}$$

where $t_0 = 0$; positive integer i takes values from 1 to τ_D, once $i = \tau_D$ then i return to 1; τ_D is the so-called "dwell time" to guarantee the stability of a slow switching control system, which is a constant depends on the upper and lower bounds of $\hat{\theta}(k)$, for details, see references [30–32].

3.2.2 Main Results

This section mainly gives two theorems as the main results. Theorem 3.1 deals with the stability and convergence of a general STC system; Theorem 3.2 deals with the stability of a general STC system.

Theorem 3.1 *If a general STC system of a deterministic plant, as described above, has the following properties:*

(1) A control strategy is well defined to stabilize the parameter known model and tracking the bounded reference signal $y_r(k)$;
(2) $\|\hat{\theta}(k)\| \leq M < \infty$; $\|\hat{\theta}(k) - \hat{\theta}(k-l)\| \to 0$, l is a finite integer;
(3) The parameter estimation error satisfies

$$\begin{aligned} e(k) &= y(k) - \phi^T(k-d)\hat{\theta}(k) \\ &= o(\alpha + \|\phi(k-d)\|), 0 < \alpha < \infty \end{aligned} \tag{3.16}$$

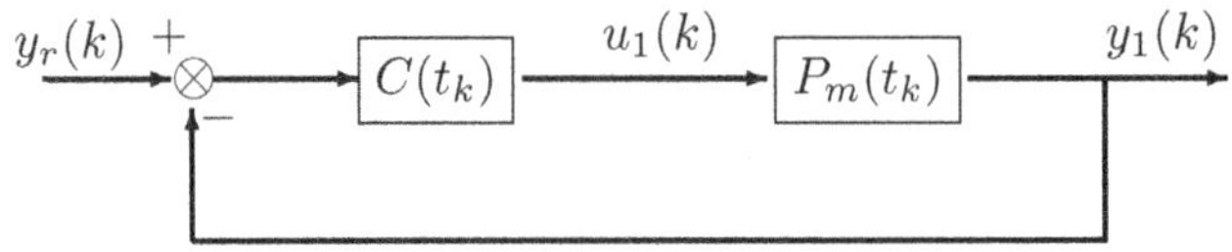

Fig. 3.5 Block diagram for subsystem 1 of deterministic VES II

where $o(.)$ is the symbol of infinitesimal, similarly hereinafter; the significance of α is to avoid division by 0.

(4) *The mapping from model parameters into controller parameters is continuous in $\hat{\theta}(k)$, for $\|\hat{\theta}(k)\| \leq M < \infty$*
Then it is stable and convergent.

Remark 3.1 Condition (2) does not imply the convergence of $\hat{\theta}(k)$; Instead of the convergence of parameter estimates, Theorem 3.1 actually requires an asymptotic property of parameter estimates, which is commonly satisfied by most existing estimation algorithms under some mild assumptions.

Proof In Fig. 3.3, we have

$$e_i(k) = y(k) - \phi^T(k-d)\hat{\theta}(t_k) = y(k) - \phi^T(k-d)\hat{\theta}(k) + \phi^T(k-d)[\hat{\theta}(k) - \hat{\theta}(t_k)] \quad (3.17)$$

This together with Conditions (2) and (3) leads to

$$e_i(k) = o(\alpha + \|\phi(k-d)\|), 0 < \alpha < \infty \quad (3.18)$$

$\|\hat{\theta}(k)\| \leq M < \infty$ in Condition (2) and Condition (4) guarantee the uniform continuity of mapping $f(.)$. Considering $\|\hat{\theta}(k) - \hat{\theta}(k-l)\| \to 0$ in Condition (2), we know that $\|\theta_c(k) - \theta_c(t_k)\| \to 0$, Then from (3.13) it is obvious that

$$\Delta u'(k) = o(\|\phi_c(k)\|) \quad (3.19)$$

Next, we decompose the VES II into three subsystems, as shown in Figs. 3.5, 3.6, and 3.7, respectively. To coincide with the initial condition of the original STC system (Fig. 3.1), we need to set $y_1(k) = 0$, $u_1(k) = 0$, $y_2(k) = 0$, $u_2(k) = 0$, $y_3(k) = 0$, $u_3(k) = 0$ for $k < 0$. For the proof of the input-output equivalence of the decomposition, see Lemma A.3 in Appendix.

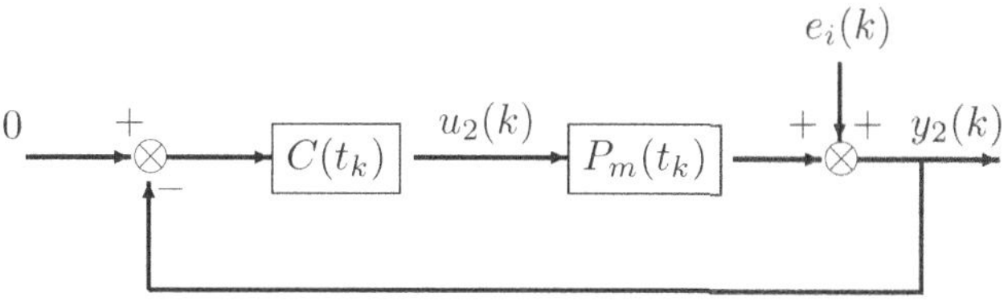

Fig. 3.6 Block diagram for subsystem 2 of deterministic VES II

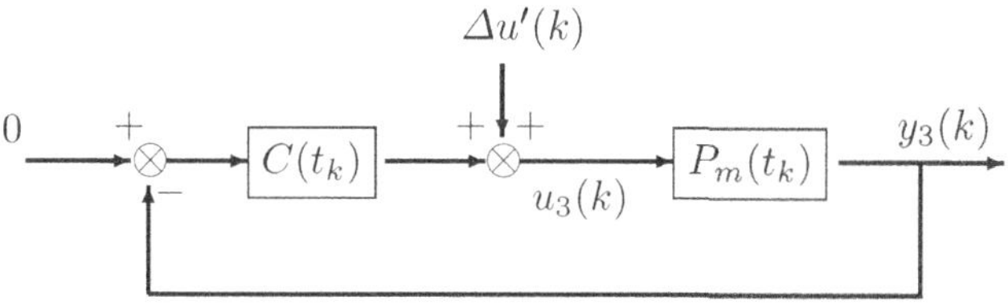

Fig. 3.7 Block diagram for subsystem 3 of deterministic VES II

According to superposition principle, we have

$$y(k) = y_1(k) + y_2(k) + y_3(k) \tag{3.20}$$

$$u(k) = u_1(k) + u_2(k) + u_3(k) \tag{3.21}$$

To facilitate the proof, we need to define a new vector $\tilde{\phi}(k)$, whose elements are the union of that of $\phi(k-d)$ and $\phi_c(k)$. Without loss of generality, we assume $s_1 < n, s_2 < m$, then $\tilde{\phi}(k)$ takes the form of

$$\tilde{\phi}(k) = [y(k), \ldots, y(k-n), u(k-1), \ldots, u(k-d-m), y_r(k), \ldots, y_r(k-s_3)] \tag{3.22}$$

Similarly, define its counterparts $\tilde{\phi}_1(k)$ in subsystem 1 (Fig. 3.5), $\tilde{\phi}_2(k)$ in subsystem 2 (Fig. 3.6), and $\tilde{\phi}_3(k)$ in subsystem 3 (Fig. 3.7), respectively, i. e.

$$\tilde{\phi}_1(k) = [y_1(k), \ldots, y_1(k-n), u_1(k-1), \ldots, u_1(k-d-m), y_r(k), \ldots, y_r(k-s_3)] \tag{3.23}$$

$$\tilde{\phi}_2(k) = [y_2(k), \ldots, y_2(k-n), u_2(k-1), \ldots, u_2(k-d-m), 0, \ldots, 0] \tag{3.24}$$

$$\tilde{\phi}_3(k) = [y_3(k), \cdots, y_3(k-n), u_3(k-1), \ldots, u_3(k-d-m), 0, \ldots, 0] \tag{3.25}$$

Then, we have

$$\tilde{\phi}(k) = \tilde{\phi}_1(k) + \tilde{\phi}_2(k) + \tilde{\phi}_3(k) \tag{3.26}$$

$$\|\phi(k-d)\| = O(\|\tilde{\phi}(k)\|) \tag{3.27}$$

$$\|\phi_c(k)\| = O(\|\tilde{\phi}(k)\|) \tag{3.28}$$

where $\|\phi(k-d)\| = O(\|\tilde{\phi}(k)\|)$ means that $\|\phi(k-d)\|/\|\tilde{\phi}(k)\| < \infty, k \to \infty$, similarly hereinafter.

From the choice of t_k, we know that all the three subsystems are deterministic "slow switching" systems with large enough dwell time. According to Condition (1) we know that for each "frozen" t_k, $C(t_k)$ and $P_m(t_k)$ constitute a stable closed-loop system. Additionally, in Condition (2), $\|\hat{\theta}(k)\| \leq M < \infty$ guarantees the index set of switching is a compact subset of a finite dimensional space. Thus, by Multiple Lyapunov Function (MLF) methodology [30–33], subsystem 1 (as shown in Fig. 3.5) is stable.

That means

$$\|\tilde{\phi}_1(k)\| < \infty \tag{3.29}$$

From the viewpoint of transfer function, we know that Subsystem 2 (Fig. 3.6) and Subsystem 3 (Fig. 3.7) are also stable, for that they have the same characteristic polynomial as Subsystem 1. The input signals of Subsystem 2 and Subsystem 3 were defined by (3.18), and (3.19), respectively. Then, taking account of (3.27) and (3.28), we obtain

$$\|y_2(k)\| = o(\alpha + \|\tilde{\phi}(k)\|) \tag{3.30}$$

$$\|u_2(k)\| = o(\alpha + \|\tilde{\phi}(k)\|) \tag{3.31}$$

$$\|\tilde{\phi}_2(k)\| = o(\alpha + \|\tilde{\phi}(k)\| + \cdots + \|\tilde{\phi}(k-l)\|) \tag{3.32}$$

$$\|y_3(k)\| = o(\|\tilde{\phi}(k)\|) \tag{3.33}$$

$$\|u_3(k)\| = o(\|\tilde{\phi}(k)\|) \tag{3.34}$$

$$\|\tilde{\phi}_3(k)\| = o(\|\tilde{\phi}(k)\| + \cdots + \|\tilde{\phi}(k-l)\|) \tag{3.35}$$

where $l = \max(n, d+m)$ is a limited integer.

Then by Lemma A.2 in Appendix A, regarding $\widetilde{\phi}_2(k)$ and $\widetilde{\phi}_3(k)$ as one variate, we conclude that $\|\tilde{\phi}(k)\|$ is bounded. It means that the STC system is stable, i.e.

$$\|y(k)\| < \infty, \quad \|u(k)\| < \infty \tag{3.36}$$

This together with (3.30), (3.31), (3.33), and (3.34), leads to

$$\|y_2(k)\| = o(1), \quad \|u_2(k)\| = o(1) \tag{3.37}$$

$$\|y_3(k)\| = o(1), \quad \|u_3(k)\| = o(1) \tag{3.38}$$

Consequently, from (3.20) and (3.21), it follows that

$$y(k) \to y_1(k) \tag{3.39}$$

$$u(k) \to u_1(k) \tag{3.40}$$

Next, we prove the tracking performance of subsystem 1 according to MLF approach, but we need to change the view-angle a little bit. Denote the tracking error of each "frozen" system $\sum\{C(\bar{t}_k), P_m(\bar{t}_k)\}$ by

$$e_{\bar{t}_k}(k) = y_{\bar{t}_k}(k) - y_r(k)$$

where $\bar{t}_k$ represents the "frozen" t_k.

Similarly, the tracking error of $\sum\{C(t_k), P_m(t_k)\}$, i.e., subsystem 1 (Fig. 3.5), is denoted by

$$e_{t_k}(k) = y_{t_k}(k) - y_r(k) = y_1(k) - y_r(k)$$

Then, for each "frozen" system $\sum\{C(\bar{t}_k), P_m(\bar{t}_k)\}$, there exists a dynamic error system

$$e_{\bar{t}_k}(k+1) = f_{\bar{t}_k}(e_{\bar{t}_k}(k), k)$$

Thus, for $\sum\{C(t_k), P_m(t_k)\}$, there exists a "slow switching" dynamic system

$$e_{t_k}(k+1) = f_{t_k}(e_{t_k}(k), k)$$

According to Condition (1), for each $\bar{t}_k$, the stable equilibrium is $e_{\bar{t}_k} = 0$, thus, there must exist a Lyapunov function $V_{\bar{t}_k}(e)$ satisfying

$$V_{\bar{t}_k}(e_{\bar{t}_k}(k)) \geq 0,\ \Delta V_{\bar{t}_k}(e_{\bar{t}_k}(k)) < 0,\ \lim_{k\to\infty} V_{\bar{t}_k}(e_{\bar{t}_k}(k)) = 0$$

Further, based on MLF methodology and "slow switching", we get a similar result for subsystem 1, i.e.

$$V(e_{t_k}(k)) \geq 0,\ \Delta V(e_{t_k}(k)) < 0,\ \lim_{k\to\infty} V(e_{t_k}(k)) = 0$$

This implies the tracking performance of the "slow switching" subsystem 1, i.e.

$$e_{t_k}(k) = y_{t_k}(k) - y_r(k) = y_1(k) - y_r(k) \to 0$$

This together with (3.39) shows that $y(k) \to y_r(k)$, i.e., the self-tung control system is convergent.

That completes the proof of Theorem 3.1. □

Further, we have the following result on the stability of a general deterministic STC system.

Theorem 3.2 *If a general STC system of a deterministic plant, as described above, has the following properties:*

(1) A control strategy is well defined to stabilize the parameter known model;
(2) $\|\hat{\theta}(k)\| \leq M < \infty$; $\|\hat{\theta}(k) - \hat{\theta}(k-l)\| \to 0$, *l is a finite integer;*
(3) The parameter estimation error is bounded, i.e.

$$\|e(k)\| = \|y(k) - \phi^T(k-d)\hat{\theta}(k)\| < \infty \tag{3.41}$$

(4) The mapping from model parameters into controller parameters is continuous in $\hat{\theta}(k)$, *for* $\|\hat{\theta}(k)\| \leq M < \infty$
Then it is stable.

Proof We still use Fig. 3.3 as the VES for this situation. The proof of Theorem 3.2 is similar to that of Theorem 3.1. The only difference is that the property of $e_i(k)$ in Fig. 3.3 has been changed. The proof is carried out in three steps.

Step 1:
The VES (as shown in Fig. 3.3) can be decomposed into three sub-systems as shown in Figs. 3.5, 3.6, and 3.7, respectively. Noticing that in Fig. 3.6 we have

$$\begin{aligned} e_i(k) &= y(k) - \phi^T(k-d)\hat{\theta}(t_k) \\ &= y(k) - \phi^T(k-d)\hat{\theta}(k) + \phi^T(k-d)[\hat{\theta}(k) - \hat{\theta}(t_k)] \\ &= e(k) + \phi^T(k-d)[\hat{\theta}(k) - \hat{\theta}(t_k)] \\ &= e(k) + e'(k) \end{aligned} \tag{3.42}$$

From Condition (3) we know $\|e(k)\| < \infty$; and from Condition (2) we have $e'(k) = o(\|\phi(k-d)\|)$.

Next we regard $e_i(k)$ as two parts, i.e., $e(k)$ and $e'(k)$. Accordingly, Fig. 3.6 is further decomposed into two sub-systems as shown in Figs. 3.8 and 3.9, respectively.

Then we see that the VES system Fig. 3.3 is actually decomposed into four sub-systems, as shown in Figs. 3.5, 3.7, 2.8, and 3.9, respectively.

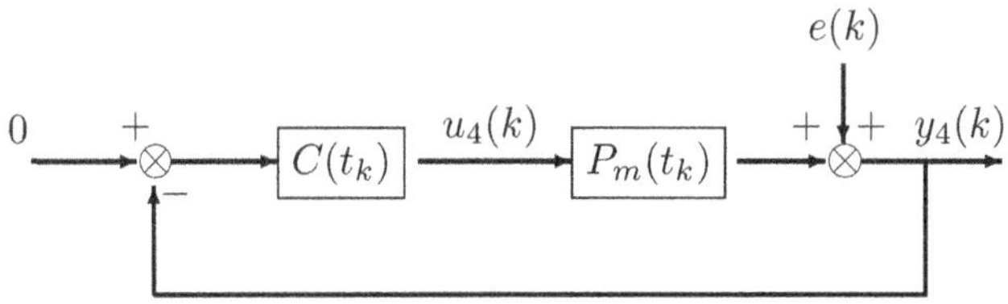

Fig. 3.8 Block diagram for subsystem 1 of system Fig. 3.6

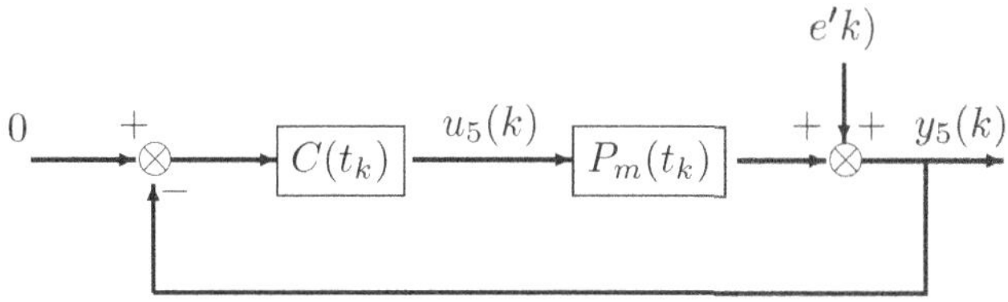

Fig. 3.9 Block diagram for subsystem 2 of system Fig. 3.6

Then we have

$$y(k) = y_1(k) + y_2(k) + y_3(k) = y_1(k) + y_3(k) + y_4(k) + y_5(k) \tag{3.43}$$

$$u(k) = u_1(k) + u_2(k) + u_3(k) = u_1(k) + u_3(k) + u_4(k) + u_5(k) \tag{3.44}$$

The properties of Figs. 3.5 and 3.7 remain the same as in the proof of Theorem 3.1.

Step 2:
Define $\tilde{\phi}(k)$, $\tilde{\phi}_1(k)$, and $\tilde{\phi}_3(k)$ the same as in the proof of Theorem 3.1. $\tilde{\phi}_4(k)$ and $\tilde{\phi}_5(k)$ can be defined similarly.

It is obvious that

$$\tilde{\phi}(k) = \tilde{\phi}_1(k) + \tilde{\phi}_3(k) + \tilde{\phi}_4(k) + \tilde{\phi}_5(k) \tag{3.45}$$

By the same reason as in the proof of Theorem 3.1, we get

$$\|\tilde{\phi}_1(k)\| < \infty \tag{3.46}$$

$$\|\tilde{\phi}_4(k)\| < \infty \tag{3.47}$$

$$\|\tilde{\phi}_3(k)\| = o(\|\tilde{\phi}(k)\| + \cdots + \|\tilde{\phi}(k-l)\|) \tag{3.48}$$

$$\|\tilde{\phi}_5(k)\| = o(\|\tilde{\phi}(k)\| + \cdots + \|\tilde{\phi}(k-l)\|) \tag{3.49}$$

Step 3:
Regarding $\tilde{\phi}_1(k)$ and $\tilde{\phi}_4(k)$ as one variate, $\tilde{\phi}_3(k)$ and $\tilde{\phi}_5(k)$ as another variate, respectively, we conclude by Lemma A.2, that $\|\tilde{\phi}(k)\|$ is bounded. That means the STC system is stable, i.e.

$$\|y(k)\| < \infty, \quad \|u(k)\| < \infty \tag{3.50}$$

The proof is completed. □

Remark 3.2 From the conditions of Theorems 3.1 and 3.2, we see that the parameter estimation algorithm and the control strategy are both changeable within limited time. Thus we suggest using order reduction (pole-zero cancelation) to tackle the uncontrollability problem of estimated model. By doing so, the properties of the four conditions of Theorems 3.1 and 3.2 remain unchanged.

3.3 Stochastic STC

3.3.1 Construction of VES

In this section, we consider a single input single output (SISO) stochastic plant P with unknown structure information and unknown parameters; it may be linear or non-linear, time-invariant or time varying, or even with unmodelled dynamics. Similar to the deterministic situations, there are no any direct limitations on the plant structure; instead some certain conditions are requested in the performances of parameter estimates. For simplicity of description, we consider stochastic plant with white noise.

Suppose the model we obtain through parameter estimation takes the form

$$P_m(k) = \sum[\hat{A}(k, q^{-1}), \hat{B}(k, q^{-1}), d, n, m] \tag{3.51}$$

where

$$\begin{aligned}\hat{A}(k, q^{-1}) &= 1 + \hat{a}_1 q^{-1} + \cdots + \hat{a}_n q^{-n} \\ \hat{B}(k, q^{-1}) &= \hat{b}_0 + \hat{b}_1 q^{-1} + \cdots + \hat{b}_m q^{-m}\end{aligned}$$

$\hat{\theta}^T(k)$ is the parametrization of $P_m(k)$

$$\hat{\theta}^T(k) = [-\hat{a}_1, \ldots, -\hat{a}_n, \hat{b}_0, \hat{b}_1, \ldots, \hat{b}_m] \tag{3.52}$$

The corresponding STC system for the above-mentioned stochastic plant P is shown in Fig. 3.10 (the estimator is omitted for simplicity), its virtual equivalent systems are shown in Figs. 3.11 and 3.12, where $y(k)$, $u(k)$ and $\omega(k)$ are the output, input, and disturbance of the plant, respectively, $y(k) = 0, u(k) = 0, \omega(k) = 0$ for $k < 0$. $y_r(k)$ is a given bounded reference signal of the system, $y_r(k) = 0$ for $k < 0$.

$C(k)$ denotes the adaptive controller, it may be designed by any existing principle, such as by pole placement. To be specific, $C(k)$ is determined by a mapping from model parameters set into controller parameters set, i.e.

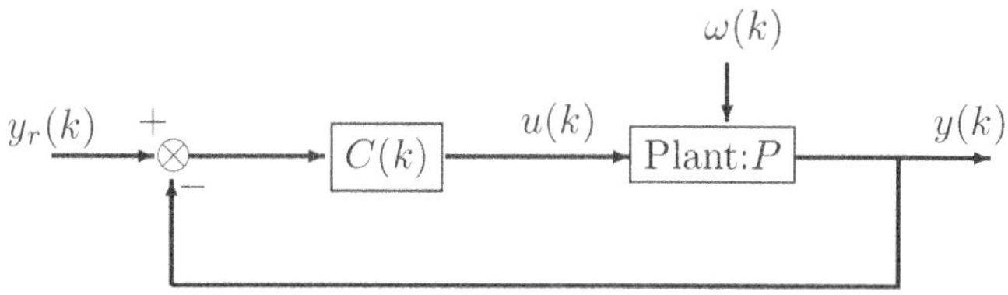

Fig. 3.10 Block diagram for stochastic self-tuning Control System

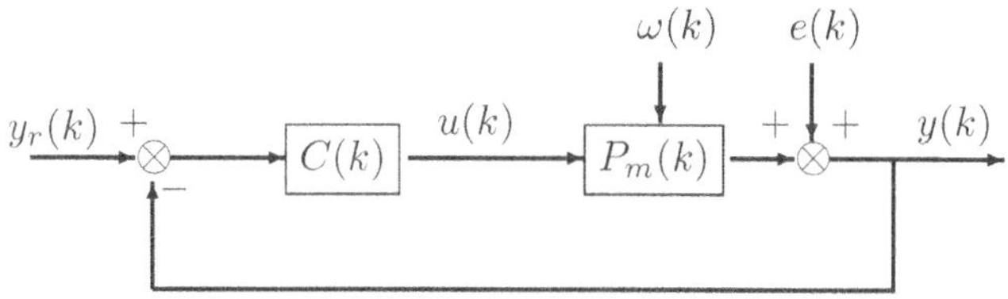

Fig. 3.11 Block diagram for stochastic VES I

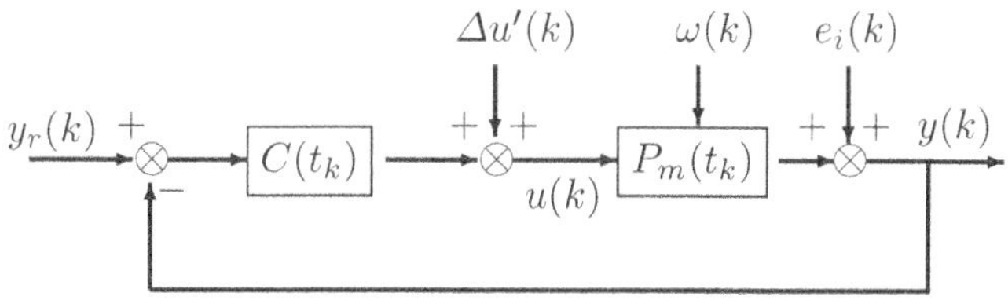

Fig. 3.12 Block diagram for stochastic VES II

$$f : \mathbb{P} \rightarrow \mathbb{C} \tag{3.53}$$

where $\mathbb{P}$ is the set of estimated models including the true model of the plant. $\mathbb{C}$ is the corresponding controllers set. Then we have

$$C(k) = f(P_m(k)) \tag{3.54}$$

or

$$\theta_c(k) = f(\hat{\theta}(k)) \tag{3.55}$$

where $\theta_c(k)$ is the parametrization of $C(k)$.

Then we have

$$u(k) = \phi_c^T(k)\theta_c(k) \tag{3.56}$$

where

$$\begin{aligned} \phi_c^T(k) &= [y(k), y(k-1), \ldots, y(k-s_1), u(k-1), \\ &\quad \ldots, u(k-s_2), y_r(k), \ldots, y_r(k-s_3)] \end{aligned} \tag{3.57}$$

The elements of $\phi_c(k)$ and the limited integers s_1, s_2, s_3 depend on specific control strategy.

A stochastic STC system as shown in Fig. 3.10 is obviously equivalent to the system as shown in Fig. 3.11, where

$$e(k) = y(k) - \phi^T(k-d)\hat{\theta}(k) - \omega(k) \tag{3.58}$$

Similar to the deterministic STC system, suppose $\|\hat{\theta}(k)\| \leq M < \infty$, $\|\hat{\theta}(k) - \hat{\theta}(k-l)\| \to 0$, l is a finite integer; the mapping $f(.)$ is uniformly continuous on the estimated parameter set. Then we can further construct a stochastic slow switching VES as shown in Fig. 3.12, where

$$\begin{aligned} e_i(k) &= y(k) - \phi^T(k-d)\hat{\theta}(t_k) - \omega(k) \\ &= y(k) - \phi^T(k-d)\hat{\theta}(k) - \omega(k) \\ &\quad +\phi^T(k-d)[\hat{\theta}(k) - \hat{\theta}(t_k)] \end{aligned} \tag{3.59}$$

$$\Delta u'(k) = \phi_c^T(k)[\theta_c(k) - \theta_c(t_k)] \tag{3.60}$$

t_k is a subsequence of k, which can be chosen the same as in deterministic situation.

Next we will develop the main results based on VES II as shown in Fig. 3.12.

3.3.2 *Main Results*

The main results are expressed by two theorems. Theorem 3.3 deals with the stability and convergence of a general STC system; Theorem 3.4 deals with the stability of a general STC system. All the limit operations in this section are in the sense of probability one.

Theorem 3.3 *If a general STC system of a stochastic plant with unknown structure information and parameters, as described above, has the following properties:*

(1) A control strategy is well defined to stabilize the parameter known model and tracking the bounded reference signal $y_r(k)$ in mean square sense;
(2) $\|\hat{\theta}(k)\| \leq M < \infty$, $\|\hat{\theta}(k) - \hat{\theta}(k-l)\| \to 0$, l is a finite integer.
(3) The parameter estimation error satisfies

$$\sum_{k=1}^{n} \|y(k) - \phi^T(k-d)\hat{\theta}(k) - \omega(k)\|^2 = o\left(1 + \sum_{k=1}^{n} \|\phi(k-d)\|^2\right) \tag{3.61}$$

(4) The mapping from model parameters into controller parameters is continuous in $\hat{\theta}(k)$, for $\|\hat{\theta}(k)\| \leq M < \infty$ Then it is stable and convergent.

Remark 3.3 Condition (2) does not imply the convergence of $\hat{\theta}(k)$; Condition (3) in Theorem 3.3 is identical to

$$\sum_{k=1}^{n} \|\hat{\omega}(k) - \omega(k)\|^2 = o\left(1 + \sum_{k=1}^{n} \|\phi(k-d)\|^2\right) \tag{3.62}$$

in most other literatures.

Proof First, according to Lemma A.3 in Appendix A (for stochastic situation, the proof procedures are the similar), VES II (Fig. 3.12) can be decomposed into three subsystems, as shown in Figs. 3.13, 3.14, and 3.15, respectively, $y_1(k) = 0$, $u_1(k) = 0$, $y_2(k) = 0$, $u_2(k) = 0$, $y_3(k) = 0$, $u_3(k) = 0$ for $k < 0$.

Then we have

$$y(k) = y_1(k) + y_2(k) + y_3(k) \tag{3.63}$$

$$u(k) = u_1(k) + u_2(k) + u_3(k) \tag{3.64}$$

To facilitate the proof, we need to define a new vector $\tilde{\phi}(k)$, whose elements are the union of that of $\phi(k-d)$ and $\phi_c(k)$. Without loss of generality, we assume $s_1 < n$, $s_2 < m$, then $\tilde{\phi}(k)$ takes the form of

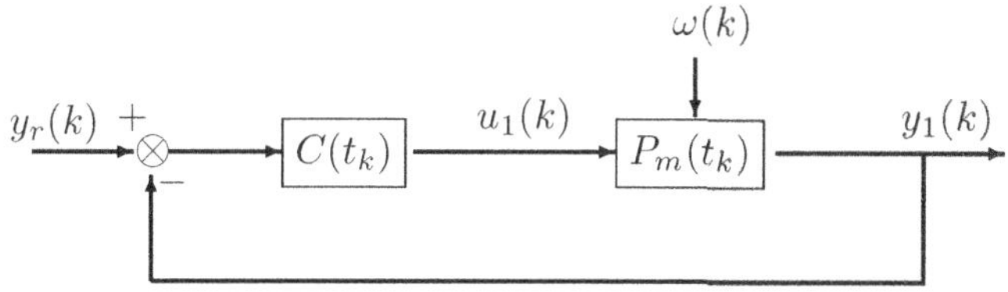

Fig. 3.13 Block diagram for subsystem 1 of stochastic VES II

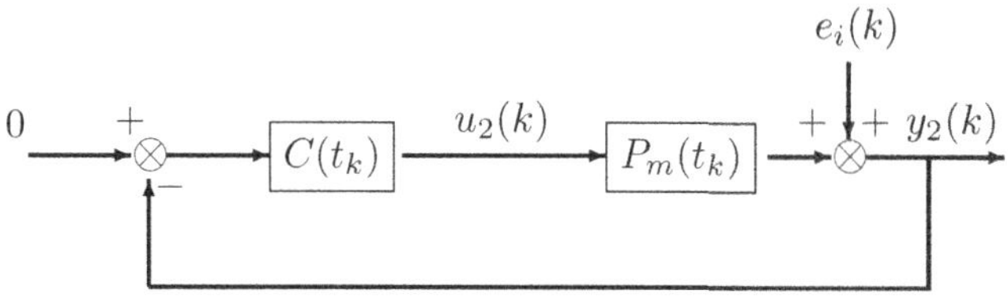

Fig. 3.14 Block diagram for subsystem 2 of stochastic VES II

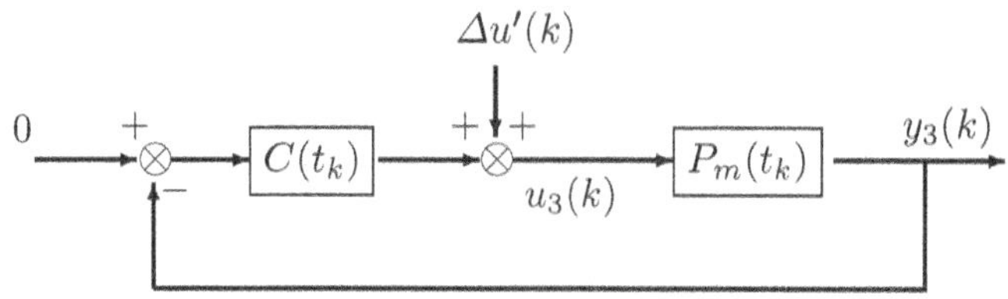

Fig. 3.15 Block diagram for subsystem 3 of stochastic VES II

$$\begin{aligned}\tilde{\phi}(k) = & \ [y(k), \ldots, y(k-n), u(k-1), \ldots, \\ & \ u(k-d-m), y_r(k), \ldots, y_r(k-s_3)]\end{aligned} \tag{3.65}$$

Similarly, define its counterparts $\tilde{\phi}_1(k)$ in subsystem 1 (Fig. 3.13), $\tilde{\phi}_2(k)$ in subsystem 2 (Fig. 3.14), and $\tilde{\phi}_3(k)$ in subsystem 3 (Fig. 3.15), respectively, i. e.

$$\begin{aligned}\tilde{\phi}_1(k) = & \ [y_1(k), \ldots, y_1(k-n), u_1(k-1), \ldots, \\ & \ u_1(k-d-m), y_r(k), \ldots, y_r(k-s_3)]\end{aligned} \tag{3.66}$$

$$\begin{aligned}\tilde{\phi}_2(k) = & \ [y_2(k), \ldots, y_2(k-n), u_2(k-1), \ldots, \\ & \ u_2(k-d-m), 0, \ldots, 0]\end{aligned} \tag{3.67}$$

$$\begin{aligned}\tilde{\phi}_3(k) = & \ [y_3(k), \ldots, y_3(k-n), u_3(k-1), \ldots, \\ & \ u_3(k-d-m), 0, \ldots, 0]\end{aligned} \tag{3.68}$$

Then we have

$$\widetilde{\phi}(k) = \widetilde{\phi}_1(k) + \widetilde{\phi}_2(k) + \widetilde{\phi}_3(k) \tag{3.69}$$

$$\|\phi_c(k)\| = O(\|\widetilde{\phi}(k)\|), \quad \|\phi(k-d)\| = O(\|\widetilde{\phi}(k)\|) \tag{3.70}$$

The choice of sequence together with Condition (2) guarantees the VES II (Fig. 3.12) is a "slow switching" system with large enough dwell time and a compact index set of switching. Then subsystem 1 (Fig. 3.13) is a stochastic slow switching system, and the other two subsystems (Figs. 3.14 and 3.15) are deterministic slow switching systems.

Further, according to Condition (1) we know that for each "frozen" t_k , $C(t_k)$ and $P_m(t_k)$ constitute a stable closed-loop system. Then, based on the existing results [34–36], we know that all the three subsystems are stable. That means in subsystem 1 (Fig. 3.13) we have

$$\frac{1}{n}\sum_{k=1}^{n}\|\widetilde{\phi}_1(k)\|^2 < \infty \tag{3.71}$$

In subsystem 2 (Fig. 3.14), with input signal given by (3.59), taking account of Conditions (2) and (3), we have

$$\begin{aligned}\sum_{k=1}^{n}\|e_i(k)\|^2 &= o\left(1+\sum_{k=1}^{n}\|\phi(k-d)\|^2\right)\\ &= o\left(1+\sum_{k=1}^{n}\|\widetilde{\phi}(k)\|^2\right)\end{aligned} \tag{3.72}$$

Then it follows that

$$\frac{1}{n}\sum_{k=1}^{n}\|y_2(k)\|^2 = o\left(\frac{1}{n}\sum_{k=1}^{n}\|\widetilde{\phi}(k)\|^2\right) \tag{3.73}$$

$$\frac{1}{n}\sum_{k=1}^{n}\|u_2(k)\|^2 = o\left(\frac{1}{n}\sum_{k=1}^{n}\|\widetilde{\phi}(k)\|^2\right) \tag{3.74}$$

$$\frac{1}{n}\sum_{k=1}^{n}\|\widetilde{\phi}_2(k)\|^2 = o\left(\frac{1}{n}\sum_{k=1}^{n}\|\widetilde{\phi}(k)\|^2\right) \tag{3.75}$$

In subsystem 3 (Fig. 3.15), by Conditions (2) and (4), and (3.60), its input signal has the following property

$$\Delta u'(k) = o(\|\phi_c(k)\|) = o(\|\widetilde{\phi}(k)\|) \tag{3.76}$$

Then we have

$$\frac{1}{n}\sum_{k=1}^{n}\|y_3(k)\|^2 = o\left(\frac{1}{n}\sum_{k=1}^{n}\|\widetilde{\phi}(k)\|^2\right) \tag{3.77}$$

$$\frac{1}{n}\sum_{k=1}^{n}\|u_3(k)\|^2 = o\left(\frac{1}{n}\sum_{k=1}^{n}\|\widetilde{\phi}(k)\|^2\right) \tag{3.78}$$

$$\frac{1}{n}\sum_{k=1}^{n}\|\widetilde{\phi}_3(k)\|^2 = o\left(\frac{1}{n}\sum_{k=1}^{n}\|\widetilde{\phi}(k)\|^2\right) \tag{3.79}$$

By Lemma A.2, regarding $\widetilde{\phi}_2(k)$ and $\widetilde{\phi}_3(k)$ as one variate, we come to the conclusion that the self-tuning control system is stable.

Next we proceed to prove the convergence of the self-tuning control system. First we examine the tracking performance of subsystem 1. Denote the tracking error of each "frozen" system $\sum\{C(\bar{t}_k), P_m(\bar{t}_k), \omega(k)\}$ by

$$e_{\bar{t}_k}(k) = y_{\bar{t}_k}(k) - y_r(k)$$

where $\bar{t}_k$ represents the "frozen" t_k.

Similarly, the tracking error of $\sum\{C(t_k), P_m(t_k), \omega(k)\}$, i.e., subsystem 1 (Fig. 3.13), is denoted by

$$e_{t_k}(k) = y_{t_k}(k) - y_r(k) = y_1(k) - y_r(k)$$

Then we know that for each "frozen" system $\sum\{C(\bar{t}_k), P_m(\bar{t}_k), \omega(k)\}$, there exists a stable dynamic system

$$e_{\bar{t}_k}(k+1) = f_{\bar{t}_k}(e_{\bar{t}_k}(k), \omega(k), k)$$

Subsequently, by MLF methodology, we know that for $\sum\{C(t_k), P_m(t_k), \omega(k)\}$, there exists a stable "slow switching" dynamic system.

$$e_{t_k}(k+1) = f_{t_k}(e_{t_k}(k), \omega(k), k)$$

According to Condition (1), for each $\bar{t}_k$, we have

$$\lim_{n\to\infty} sup \frac{1}{n} \sum_{k=1}^{n} \|e_{\bar{t}_k}(k)\|^2 = R$$

where R is a constant. It is easy to see that $e_{t_k}(k)$ takes values in the set of $e_{\bar{t}_k}(k)$, then by squeezing, it follows that

$$\lim_{n\to\infty} sup \frac{1}{n} \sum_{k=1}^{n} \|e_{t_k}(k)\|^2 = R$$

That implies the "slow switching" subsystem 1 is tracking, i.e.

$$\lim_{n\to\infty} sup \frac{1}{n} \sum_{k=1}^{n} \|y_1(k) - y_r(k)\|^2 = R$$

Next, by Lemma A.4, regarding $y_2(k)$ and $y_3(k)$ as one variate, we see that the self-tung control system is convergent, i.e.,

$$\begin{aligned} &\lim_{n\to\infty} sup \tfrac{1}{n} \textstyle\sum_{k=1}^{n} \|y(k) - y_r(k)\|^2 \\ &= \lim_{n\to\infty} sup \tfrac{1}{n} \textstyle\sum_{k=1}^{n} \|y_1(k) - y_r(k) + y_2(k) + y_3(k)\|^2 \\ &= R \end{aligned} \tag{3.80}$$

That completes the proof of Theorem 3.3. □

Theorem 3.4 *If a general STC system of an arbitrary stochastic plant (time-invariant or time varying, linear or nonlinear), as described above, has the following properties:*

(1) A control strategy is well defined to stabilize the parameter known model and tracking the bounded reference signal $y_r(k)$ in mean square sense;

(2) $\|\hat{\theta}(k)\| \leq M < \infty$, $\|\hat{\theta}(k) - \hat{\theta}(k-l)\| \to 0$, *$l$ is a finite integer.*

(3) The parameter estimation error is bounded, i.e.

$$\frac{1}{n}\sum_{k=1}^{n}\|y(k) - \phi^T(k-d)\hat{\theta}(k) - \omega(k)\|^2 < \infty \tag{3.81}$$

(4) The mapping from model parameters into controller parameters is continuous in $\hat{\theta}(k)$, for $\|\hat{\theta}(k)\| \leq M < \infty$ Then it is stable.

Proof We still use Fig. 3.12 as the VES in this situation. The proof of Theorem 3.4 is similar to that of Theorem 3.3. The only difference is that the property of $e_i(k)$ in Fig. 3.12 has been changed. The proof is carried out in three steps.

Step 1:
Decompose the VES system Fig. 3.12 into three subsystems as shown in Figs. 3.13, 3.14, and 3.15, respectively. Notice that in Fig. 3.14 we have

$$\begin{aligned} e_i(k) &= y(k) - \phi^T(k-d)\hat{\theta}(t_k) - \omega(k) \\ &= y(k) - \phi^T(k-d)\hat{\theta}(k) - \omega(k) + \phi^T(k-d)[\hat{\theta}(k) - \hat{\theta}(t_k)] \\ &= e(k) + \phi^T(k-d)[\hat{\theta}(k) - \hat{\theta}(t_k)] \\ &= e(k) + e'(k) \end{aligned} \tag{3.82}$$

From Condition (3) we know

$$\frac{1}{n}\sum_{k=1}^{n}\|e(k)\|^2 = \frac{1}{n}\sum_{k=1}^{n}\|y(k) - \phi^T(k-d)\hat{\theta}(k) - \omega(k)\|^2 < \infty \tag{3.83}$$

and from Condition (2) we have

$$e'(k) = o(\|\phi(k-d)\|) \tag{3.84}$$

Next we regard $e_i(k)$ as two parts, i.e., $e(k)$ and $e'(k)$. Accordingly, deterministic system Fig. 3.14 is further decomposed into two subsystems as shown in Figs. 3.16 and 3.17, respectively.

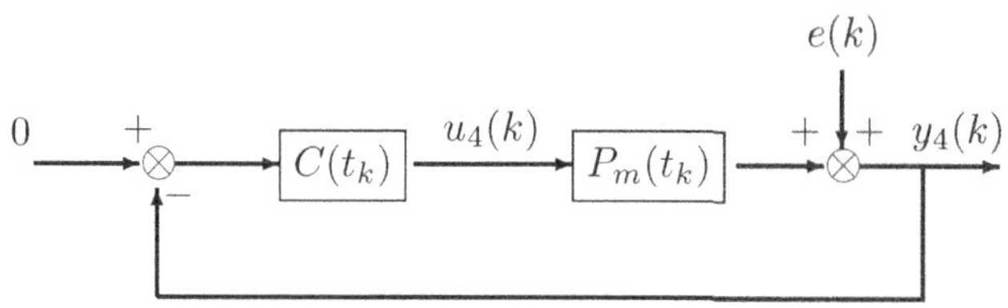

Fig. 3.16 Block diagram for subsystem 1 of System Fig. 3.14

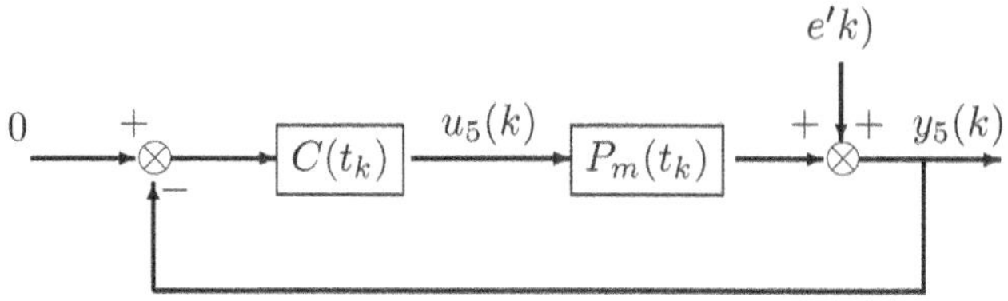

Fig. 3.17 Block diagram for subsystem 2 of System Fig. 3.14

Then we see that the VES system Fig. 3.12 is actually decomposed into four subsystems, as shown in Figs. 3.13, 3.15, 3.16, and 3.17, respectively.
Then we have

$$y(k) = y_1(k) + y_2(k) + y_3(k) = y_1(k) + y_3(k) + y_4(k) + y_5(k) \tag{3.85}$$

$$u(k) = u_1(k) + u_2(k) + u_3(k) = u_1(k) + u_3(k) + u_4(k) + u_5(k) \tag{3.86}$$

The properties of Figs. 3.13 and 3.15 remain the same as in the proof of Theorem 3.3.

Step 2:
Define $\tilde{\phi}(k)$, $\tilde{\phi}_1(k)$, and $\tilde{\phi}_3(k)$ the same as in the proof of Theorem 3.3. $\tilde{\phi}_4(k)$ and $\tilde{\phi}_5(k)$ can be defined similarly.

It is obvious that

$$\tilde{\phi}(k) = \tilde{\phi}_1(k) + \tilde{\phi}_3(k) + \tilde{\phi}_4(k) + \tilde{\phi}_5(k) \tag{3.87}$$

By the same reason as in the proof of Theorem 3.3, we get

$$\frac{1}{n}\sum_{k=1}^{n}\|\tilde{\phi}_1(k)\|^2 < \infty \tag{3.88}$$

$$\frac{1}{n}\sum_{k=1}^{n}\|\tilde{\phi}_4(k)\|^2 < \infty \tag{3.89}$$

$$\frac{1}{n}\sum_{k=1}^{n}\|\tilde{\phi}_3(k)\|^2 = o\left(\frac{1}{n}\sum_{k=1}^{n}\|\tilde{\phi}(k)\|^2\right) \tag{3.90}$$

$$\frac{1}{n}\sum_{k=1}^{n}\|\tilde{\phi}_5(k)\|^2 = o\left(\frac{1}{n}\sum_{k=1}^{n}\|\tilde{\phi}(k)\|^2\right) \tag{3.91}$$

Step 3:
Regarding $\tilde{\phi}_1(k)$ and $\tilde{\phi}_4(k)$ as one variate, $\tilde{\phi}_3(k)$ and $\tilde{\phi}_5(k)$ as another variate, respectively, we conclude by Lemma A.2, that

$$\frac{1}{n}\sum_{k=1}^{n}\|\tilde{\phi}(k)\|^2 < \infty \tag{3.92}$$

Then we know the stochastic STC system as shown in Fig. 3.10 is stable, i.e.

$$\frac{1}{n}\sum_{k=1}^{n}\|y(k)\|^2 < \infty \tag{3.93}$$

$$\frac{1}{n}\sum_{k=1}^{n}\|u(k)\|^2 < \infty \tag{3.94}$$

That completes the proof of Theorem 3.4. □

Remark 3.4 From the conditions of Theorems 3.3 and 3.4, we see that the parameter estimation algorithm and the control strategy are both changeable within limited time. Thus we suggest using order reduction (pole-zero cancelation) to tackle the uncontrollability problem of estimated model. By doing so, the properties of the four conditions of Theorems 3.3 and 3.4 remain unchanged.

3.4 Conclusion

The necessary conditions on global stability and convergence of a general STC system for both deterministic and stochastic plants have been further relaxed (compared with the results of Chap. 2), i.e., the structure information of the plant and the convergence of parameter estimates have been removed. It is therefore argued that persistent excitation is not necessary for STC systems. In other words, feedback information of the adaptive control system itself is sufficient to achieve the control objective (regulation or tracking). These perspectives provide us with more confidence and flexibility in adaptive control engineering. In addition, it is well-founded that STC system based on suitable low order model may also achieve the desired control objective. Many simulation results coincide with these observations.

Although we only considered SISO control systems, it is straightforward to extend all the results to MIMO (multiple input multiple output) systems.

References

1. R.E. Kalman, Design of a self optimizing control system. Trans. ASME **80**, 468–478 (1958)
2. K.J. Åström, B. Wittenmark, On self-tuning regulators. Automatica **9**, 185–199 (1973)
3. C.J. Harris, S.A. Billings, *Self-Tuning and Adaptive Control: Theory and Applications* (Peter Peregrinus Ltd., 1981)
4. G.C. Goodwin, K.S. Sin, *Adaptive Filtering, Prediction and Control* (Prentice Hall, 1984)
5. B.D.O. Anderson, R.R. Bitmead, C.R. Johnson Jr., P.V. Kokotovic, R.L. Kosut, I. Mareels, L. Praly, B.D. Riedle, *Stability of Adaptive Systems* (MIT Press, Cambridge, MA, 1986)
6. K.S. Narendra, A.M. Annaswamy, *Stable Adaptive Systems* (Prentice-Hall, Englewood Cliffs, 1989)
7. S. Sastry, M. Bodson, *Adaptive Control: Stability, Convergence and Robustness* (Prentice Hall, Englewood Cliffs, New Jersey, 1989)
8. K.J. Åström, B. Wittenmark, *Adaptive Control* (Addison-Wesley, 1995)
9. P.A. Ioannou, J. Sun, *Robust Adaptive Control* (Prentice-Hall: Englewood Cliffs, USA, New Jersey, 1996)
10. I.D. Landau, R. Lozano, M. M'Saad, *Adaptive Control* (Springer, London, 1998)
11. G. Tao, *Adaptive Control Design and Analysis* (Wiley, New York, 2003)
12. G.C. Goodwin, P.J. Ramadge, P.E. Caines, Discrete time stochastic adaptive control. SIAM J. Control Optim. **19**, 829–853 (1981)
13. L. Guo, H.F. Chen, The Astrom-Wittenmark self-tuning regulator revisited and ELS-based adaptive trackers. IEEE Trans. Autom. Control **36**, 802–812 (1991)
14. D.W. Clarke, P.J. Gawthrop, Self-tuning controller. Proc. IEE **122**(9), 929–934 (1975)
15. P.E. Wellstead, J.M. Edmunds, D. Prager, P. Zanker, Self-tuning pole/zero assignment regulators. Int. J. Control **30**(1), 1–26 (1979)
16. P.J. Gawthrop, On the stability and convergence of a self-tuning controller. Int. J. Control **31**(5), 973–998 (1980)
17. D.W. Clarke, Self-tuning control of non-minimum-phase systems. Automatica **20**(5), 501–517 (1984)
18. F. Giri, M. M'Saad, L. Dugard, J.M. Dion, Robust pole placement indirect adaptive control. Int. J. Adapt. Control Signal Process. **2**, 33–47 (1988)
19. G. Kreisselmeier, An indirect adaptive controller with a self-excitation capability. IEEE Trans. Autom. Control **34**, 524–528 (1989)
20. P.R. Kumar, Convergence of adaptive control schemes using least squares parameter estimates. IEEE Trans. Autom. Control **35**(4), 416–424 (1990)
21. B. Egardt, Unification of some discrete time adaptive control schemes. IEEE Trans. Autom. Control **25**, 693–697 (1980)
22. R. Lozano, X.H. Zhao, Adaptive pole placement without excitation probing signals. IEEE Trans. Autom. Control **3**(9), 47–58 (1994)
23. L. Guo, Self-convergence of weighted least—squares with applications to stochastic adaptive control. IEEE Trans. Autom. Control **41**(1), 79–89 (1996)
24. K. Nassiri-Toussi, W. Ren, Indirect adaptive pole-placement control of MIMO stochastic systems: self-tuning results. IEEE Trans. Autom. Control **42**(1), 38–52 (1997)
25. M. Prandini, M.C. Campi, Adaptive LQG control of input-output systems—acost-biased approach. SIAM J. Control Optim. **39**, 1499–1519 (2001)
26. A. Patete1, K. Furuta, M. Tomizuka, Stability of self-tuning control based on Lyapunov function. Int. J. Adapt. Control Signal Process **22**, 795–810 (2008)
27. G.C. Goodwin, A. Feuer, Adaptive control: where to now? adaptive systems for signal processing, communications, and control symposium, in *Proceedings of IEEE* (2000), pp. 165–170
28. S. Fekri, M. Athans, A. Pascoal, Issues, progress and new results in robust adaptive control. Int. J. Adapt. Control Signal Process **20**(10), 519–579 (2006)
29. W. Zhang, On the stability and convergence of self-tuning control–virtual equivalent system approach. Int. J. Control **83**(5), 879–896 (2010)

30. R. Shorten et al., Stability criteria for switched and hybrid systems. SIAM Rev. **49**, 545–592 (2007)
31. D. Liberzon, A.S. Morse, Basic problems in stability and design of switched systems. IEEE Control Syst. Mag. **19**, 59–70 (1999)
32. J.H. Frommer, S.R. Kulkarni, P.J. Ramadge, Controller switching based on output prediction errors. IEEE Trans. Autom. Control **43**, 596–607 (1998)
33. D. Chatterjee, D. Liberzon, Stability analysis of deterministic and stochastic switched systems via a comparison principle and multiple Lyapunov functions. SIAM J. Control Optim. **45**(1), 174–206 (2006)
34. D. Chatterjee, D. Liberzon, On stability of stochastic switched systems, in *Proceedings of 43rd Conference on Decision and Control*, vol. 4 (2004), pp. 4125–4127
35. M.C. Campi, J.P. Hespanha, M. Prandini, Cautious hierarchical switching control of stochastic linear systems. Int. J. Adapt. Control Signal Process. **18**, 319–333 (2004)
36. M. Prandini, Switching control of stochastic linear systems: Stability and performance results, in *Proceedings 6th Congress of SIMAI* (Chia Laguna, Cagliari, Italy, 2002)

Chapter 4
Stable Weighted Multiple Model Adaptive Control of Discrete-Time Stochastic Plant

Abstract From this chapter, we try to use VES as a general stability analysis tool for various model based control systems, such as weighted multiple model adaptive control (WMMAC), T-S model based fuzzy control. In this chapter, a stable WMMAC control scheme for uncertain linear, discrete-time stochastic plant is presented. First, a new weighting algorithm is proposed with convergence analysis, i.e., one weight (corresponding to the true model or the closest model to the plant) converges to 1, and other weights converge to 0. Second, based on VES concept and methodology, the stability of the proposed WMMAC system is proved under a unified framework which is independent of specific weighting algorithm and specific 'local' control strategy. Finally, some simulation results are presented to verify the effectiveness of the theoretical analysis results of the proposed WMMAC scheme.

4.1 Introduction

To address some difficult issues of STC, such as bad transient performance, some pioneering researchers (for example, K. S. Narendra, A. S. Morse) in adaptive control field turn to multiple model adaptive control (MMAC), including switching multiple model adaptive control and weighted multiple model adaptive control in 1990s. In fact, the research of multiple model methodology appeared in mid 1960s and early 1970s, where multiple Kalman filter (KF) based models were used to improve the accuracy of state estimation in adaptive filtering and adaptive control [1–3]. Later on, many practical applications of multiple model were reported [4–7]. A WMMAC system is always produced as the probability-weighted sum of elemental controller outputs. In recent years, a new version of WMMAC, i.e., robust multiple model adaptive control (RMMAC) scheme was put forward with experimental verifications [8–10]. As theoretical progress of WMMAC in early stage, some convergence results on the probabilistic weighting algorithm have been obtained under suitable assumptions [11–14]. Specifically, as Fekri, Athans and Pascoal pointed out in [9] that under certain ergodicity and stationarity assumptions, one of the posterior probabilities will converge almost surely to unity and will 'identify' the model closest to the true plant, i.e. the one with smallest Baram Proximity Measure (BPM).

W. Zhang and Q. Li, *Virtual Equivalent System Approach for Stability Analysis of Model-based Control Systems*, https://doi.org/10.1007/978-981-15-5538-1_4

Despite of decades of theoretical and experimental research, it is still widely accepted that the stability of WMMAC (including RMMAC) systems is rather difficult to prove [9, 15, 16]. Actually, to the best of the authors' knowledge, reference [16] seems to be the first attempt to deal with the stability of RMMAC/WMMAC, in which the weighting algorithm was modified with dwell-time switching structure.

We will make efforts, in this chapter, from two directions to prove the stability issue of discrete-time WMMAC. First, a new weighting algorithm is proposed with assured convergence under smooth conditions; Second, VES approach is adopted to address the stability of the closed-loop WMMAC system. VES is an artificial system that is equivalent to the original adaptive control system in the input-output sense. It was originated from [17] and gradually adapted later in [18–20]. VES has been successful in understanding and judgment of the stability and convergence of a general self-tuning control (STC) system, which consists of arbitrary control strategy, arbitrary parameter estimation algorithm, and a deterministic/stochastic minimum/non-minimum phase linear time-invariant (LTI) plant. For more details, the reader is referred to [20].

Besides, the proposed stability analysis method based on VES is independent of specific 'local' control strategy and weighting algorithm, instead it only requires the properties of 'local' control strategy (stabilizing and tracking) and weighting algorithm (convergence). It should be pointed out that we focused on the discrete-time WMMAC system because most practical systems are controlled by computers which are discrete in nature. As Narendra pointed out in [21], the presence of random noise can be dealt with more easily in the case of discrete-time systems. Since most practical systems have to operate in the presence of noise, the stability and performance of MMAC systems in such contexts has to be well understood, if the theory is to find wide application in practice.

To avoid ambiguity, we need to emphasize that the stability of WMMAC means the boundedness of its input-output signals; and the convergence of its performance index to that of 'local' non-adaptive control systems. In addition, we need to point out that all the limit operations in this chapter are in the sense of probability one.

4.2 Description of WMMAC System

Consider the following discrete-time stochastic single-input and single-output (SISO) plant P

$$A(q^{-1})y(k) = q^{-d}B(q^{-1})u(k) + \omega(k) \tag{4.1}$$

where

$$A(q^{-1}) = 1 + a_1q^{-1} + \cdots + a_nq^{-n}$$
$$B(q^{-1}) = b_0 + b_1q^{-1} + \cdots + b_mq^{-m}$$

$y(k), u(k)$ and $\omega(k)$ are the output, input and the exogenous disturbance/measurement noise of the system, respectively, $y(k) = 0, u(k) = 0, \omega(k) = 0$ for $k < 0$. Further we suppose $\omega(k)$ is a zero-mean white noise with constant variance $0 < R < \infty$, i.e.,

$$\lim_{k\to\infty} \frac{1}{k} \sum_{i=1}^{k} \omega^2(i) = R \tag{4.2}$$

The plant P can be stable or non-stable, minimum phase or non-minimum phase. The plant output $y(k)$ can be rewritten as

$$y(k) = \phi^T(k-d)\theta + \omega(k) \tag{4.3}$$

where

$$\phi^T(k-d) = [y(k-1), \ldots, y(k-n), u(k-d), \ldots, u(k-d-m)] \tag{4.4}$$

$$\theta = [-a_1, \ldots, -a_n, b_0, b_1, \ldots, b_m] \tag{4.5}$$

$\mathbb{M} = \{M_i, i = 1, 2, \ldots, N\}$ is the model set that may include the true model of the unknown plant P.

For each local model $M_i \in \mathbb{M}$, its output is given by

$$y_{mi}(k) = \phi^T(k-d)\theta_i \tag{4.6}$$

where θ_i is the parameter vector of model M_i. Further, define the output error of each local model M_i, i.e.

$$e_i(k) = y(k) - y_{mi}(k) = y(k) - \phi^T(k-d)\theta_i = \phi^T(k-d)\theta - \phi^T(k-d)\theta_i + \omega(k) \tag{4.7}$$

As we will see later, $e_i(k)$ is used to calculate the weight for each 'local' controller C_i, which may be designed according to any possible control strategies, if only C_i stabilizes model $M_i \in \mathbb{M}$ and the resulting closed-loop system tracks the reference input $0 < y_r(k) < \infty$. $\mathbb{C} = \{C_i, i = 1, 2, \ldots, N\}$ is the controller set corresponding to the model set $\mathbb{M} = \{M_i, i = 1, 2, \ldots, N\}$. We use a concise block diagram as shown in Fig. 4.1 to represent a general WMMAC system of discrete-time plant, in which the details of design strategy of 'local' controllers and algorithm for calculating controllers' weights $p_i(k)$ are omitted.

In Fig. 4.1, each 'local' controller C_i generates 'local' control signal $u_i(k)$, and the global control law $u(k)$ is obtained by

$$u(k) = \sum_{i=1}^{N} p_i(k)u_i(k) \tag{4.8}$$

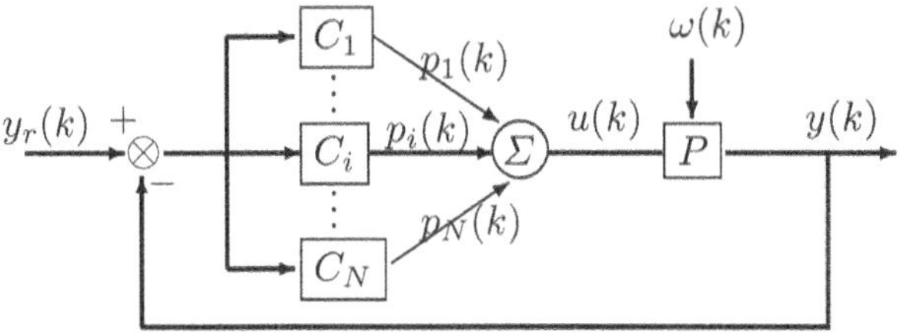

Fig. 4.1 Simplified block diagram of a general WMMAC system

Typically, controller weights are calculated based on a bank of KFs and the so-called posterior probability evaluator (PPE), for details, the reader is referred to reference [9]. But in this chapter we propose a different weighting algorithm to calculate $p_i(k)$, i.e.,

$$l_i(0) = \frac{1}{N};\; p_i(0) = l_i(0) \tag{4.9}$$

for all $k > 0$

$$l_i'(k) = 1 + \frac{1}{k}\sum_{r=1}^{k} e_i^2(r) \tag{4.10}$$

$$l_{min}'(k) = \min_i\{l_i'(k)\} \tag{4.11}$$

$$l_i(k) = \frac{l_{min}'(k)}{l_i'(k)} l_i(k-1) \tag{4.12}$$

$$p_i(k) = \frac{l_i(k)}{\sum_{r=1}^{N} l_r(k)} \tag{4.13}$$

In contrast to other existing WMMAC schemes, such as, RMMAC and classical WMMAC [9], this newly-proposed algorithm is simpler in calculation and easy to implement. We have the following convergence analysis result of the proposed weighting algorithm.

Theorem 4.1 *Suppose M_j is the closest model in the model set $\mathbb{M} = \{M_i, i = 1, 2, \ldots, N\}$ to the true plant in the following sense with probability one*

$$\frac{1}{k}\sum_{r=1}^{k} e_j^2(r) < \frac{1}{k}\sum_{r=1}^{k} e_i^2(r), \forall k \geq k^*, i \neq j \tag{4.14}$$

$$\lim_{k\to\infty} \frac{1}{k}\sum_{r=1}^{k} e_j^2(r) = R_j, \lim_{k\to\infty} \frac{1}{k}\sum_{r=1}^{k} e_i^2(r) = R_i, R_j < R_i, i \neq j \tag{4.15}$$

where k^* is an unknown limited time instant, R_j is a constant, R_i may be constant or infinity.

Then we have

$$\lim_{k\to\infty} p_j(k) = 1;\ \lim_{k\to\infty} p_i(k) = 0, i \neq j$$

Proof It is not difficult to see that algorithms (4.9)–(4.13) together with (4.14) guarantee with probability one that

$$\begin{cases} l'_{min}(k) = l'_j(k) \\ \frac{l'_{min}(k)}{l'_j(k)} = 1 \\ \frac{l'_{min}(k)}{l'_i(k)} < 1 \end{cases} \quad \forall k \geq k^*, i \neq j \tag{4.16}$$

Further, considering (4.15), we have

$$\lim_{k\to\infty} \frac{l'_{min}(k)}{l'_i(k)} = \frac{1+R_j}{1+R_i} < 1, i \neq j \tag{4.17}$$

Putting (4.16), (4.17), and (4.12) together, we get

$$\lim_{k\to\infty} l_j(k) = l_j(k^*) > 0;\ \lim_{k\to\infty} l_i(k) = 0, i \neq j \tag{4.18}$$

Then from(4.13) we obtain

$$\lim_{k\to\infty} p_j(k) = 1;\ \lim_{k\to\infty} p_i(k) = 0, i \neq j \tag{4.19}$$

That completes the proof of Theorem 4.1.

Next we discuss the relationship between the convergence conditions in Theorem 4.1 and the signal-to-noise ratio.

Considering (4.7) and that $\omega(k)$ is a white noise, we know that $e_i(k)$ consists of two independent components, the noise $\omega(k)$ and the determinant output error between the plant and the ith model, i.e., $\Delta y_i(k) = \phi^T(k-d)\theta - \phi^T(k-d)\theta_i$. Then (4.7) can be rewritten as

$$e_i(k) = \Delta y_i(k) + \omega(k) \tag{4.20}$$

Consequently, it is not difficult to imagine that the convergence conditions of weighting algorithms (4.9)–(4.13) depend on the noise power, as well as the signal power of $\Delta y_i(k)$. In other words, the index (4.10) should be discriminable under the disturbance of noise.

To be specific, let's first consider a simple situation that the true model of plant, M_j, is included in the model set, i.e., $M_j \in \mathbb{M}$. Then from (4.7) we have

$$e_j(k) = \omega(k) \tag{4.21}$$

$$l'_{min}(k) = l'_j(k) = 1 + \frac{1}{k}\sum_{r=1}^{k} e_j^2(r) \tag{4.22}$$

$$\lim_{k\to\infty}\left[1 + \frac{1}{k}\sum_{r=1}^{k} e_j^2(r)\right] = 1 + R \tag{4.23}$$

To ensure the convergence (rate) of the weighting algorithms (4.9)–(4.13), we need, with probability one, that

$$\frac{l'_{min}(k)}{l'_i(k)} \le \frac{1}{K}, K > 1, k \ge k^* \tag{4.24}$$

i.e.,

$$l'_i(k) \ge K * l'_j(k), K > 1, k \ge k^*, i \ne j \tag{4.25}$$

This together with (4.10) and (4.20) yields

$$\lim_{k\to\infty}\left[1 + \frac{1}{k}\sum_{r=1}^{k} e_i^2(r)\right] \ge K * \lim_{k\to\infty}\left[1 + \frac{1}{k}\sum_{r=1}^{k} e_j^2(r)\right], i \ne j \tag{4.26}$$

Further, considering that $\Delta y_i(k)$ and $\omega(k)$ are independent, we have

$$1 + R + \lim_{k\to\infty}\frac{1}{k}\sum_{r=1}^{k}[\Delta y_i(r)]^2 \ge K * (1 + R) \tag{4.27}$$

Denote $P_{\Delta y_i} = \lim_{k\to\infty}\frac{1}{k}\sum_{r=1}^{k}[\Delta y_i(r)]^2$, then we obtain

$$\frac{P_{\Delta y_i}}{1 + R} \ge K - 1 \tag{4.28}$$

Equation (4.28) implies that if we want sharper convergence rate, then we need higher signal-to-noise ratio, i.e., $P_{\Delta y_i}/(1 + R)$, which, considering R is a constant, depends on the difference between the true model of plant and each of the other models.

Actually, (4.28) represents an upper bound for K, i.e., $K \le 1 + \frac{P_{\Delta y_i}}{1+R}$, while a lower bound for K could be more useful, i.e.,

$$K \ge 1 + \alpha \tag{4.29}$$

where α should be decided by word length and rounding (truncating) rules of the floating-point system to avoid that $1/K$ is approximately 1.

Similarly, for the situation that the true model of plant is not included in the model set, but $M_j \in \mathbb{M}$ is the closest one to the true model of plant, we have the following limitations on K

$$\frac{1 + R + P_{\Delta y_i}}{1 + R + P_{\Delta y_j}} \geq K \geq 1 + \alpha, i \neq j \tag{4.30}$$

where $P_{\Delta y_j} = \lim_{k\to\infty} \frac{1}{k} \sum_{r=1}^{k} [\Delta y_j(r)]^2$, and α is the same as in (4.29).

4.3 Virtual Equivalent System

This section describes two types of VESs of WMMAC under the condition that $\lim_{k\to\infty} p_j(k) = 1$; $\lim_{k\to\infty} p_i(k) = 0, i \neq j$, where j indicates the closest model $M_j \in \mathbb{M}$ to the true plant. For the first type of VES, M_j is the true model of plant; for the second type of VES, M_j is not the true model of plant.

4.3.1 Type I of VES

Suppose

$$\lim_{k\to\infty} p_j(k) = 1; \lim_{k\to\infty} p_i(k) = 0, i = 1, \ldots, N, i \neq j \tag{4.31}$$

M_j is the true model of the plant P.

Then we have a virtual equivalent system of WMMAC based on $C = C_j$ and $P = M_j$, as shown in Fig. 4.2, where

$$u(k) = \sum_{i=1}^{N} p_i(k) \cdot u_i(k) \tag{4.32}$$

$u_i(k)$ is the output of 'local' controller $C_i, i = 1, \ldots, j, \ldots, N$, $\Delta u_j(k)$ is the controller output difference between global control $u(k)$ and local control $u_j(k)$, i.e.

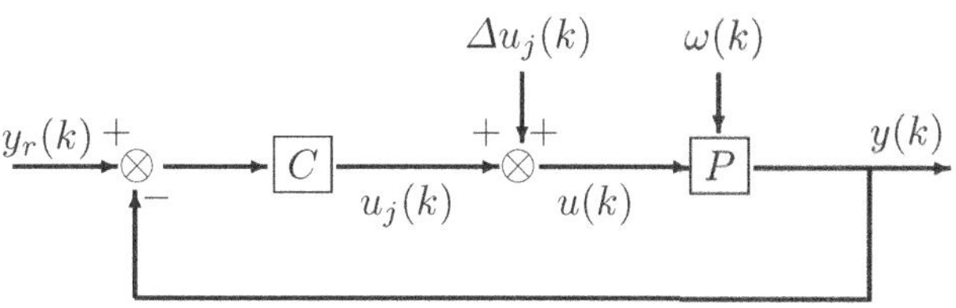

Fig. 4.2 Virtual equivalent system I of a WMMAC system

$$\begin{aligned}\Delta u_j(k) &= u(k) - u_j(k)\\ &= \sum_{i=1}^{N} p_i(k)\cdot u_i(k) - u_j(k)\\ &= [(p_j(k)-1]\cdot u_j(k) + \sum_{i=1,i\neq j}^{N} p_i(k)\cdot u_i(k)\end{aligned} \tag{4.33}$$

Without loss of generality, denote

$$u_i(k) = \phi_c^T(k)\theta_{ci} \tag{4.34}$$

where

$$\begin{aligned}\phi_c^T(k) = [&y(k), y(k-1), \ldots, y(k-s_1), u(k-1),\\ &\ldots, u(k-s_2), y_r(k), \ldots, y_r(k-s_3)]\end{aligned} \tag{4.35}$$

is the regression vector form of local control signal $u_i(k)$. The numbers of the elements of $\phi_c^T(k)$, i.e., s_1, s_2, s_3, are limited integers and depend on specific local control strategy; θ_{ci} is the parameter vector form of 'local' controller C_i.

Putting (4.3) and (4.4) together, we obtain

$$\frac{\Delta u_j(k)}{\|\phi_c(k)\|} = [p_j(k)-1]\frac{\phi_c^T(k)\theta_{cj}}{\|\phi_c(k)\|} + \sum_{i=1,i\neq j}^{N} p_i(k)\frac{\phi_c^T(k)\theta_{ci}}{\|\phi_c(k)\|} \tag{4.36}$$

Considering (4.1) and that $\theta_{ci}, i = 1, \ldots, j, \ldots, N$ are constant vectors, we have

$$\lim_{k\to\infty}\frac{\Delta u_j(k)}{\|\phi_c(k)\|} = 0 \tag{4.37}$$

i.e.,

$$\Delta u_j(k) = o(\|\phi_c(k)\|) \tag{4.38}$$

The Little-Oh operator is defined in Appendix.

As we will see in the next section, the property of $\Delta u_j(k)$ together with the stabilizing and tracking characteristics of 'local' control strategy is the key factor to the stability of the proposed WMMAC system.

4.3.2 *Type II of VES*

Suppose

$$\lim_{k\to\infty} p_j(k) = 1;\ \lim_{k\to\infty} p_i(k) = 0, i = 1, \ldots N, i \neq j \tag{4.39}$$

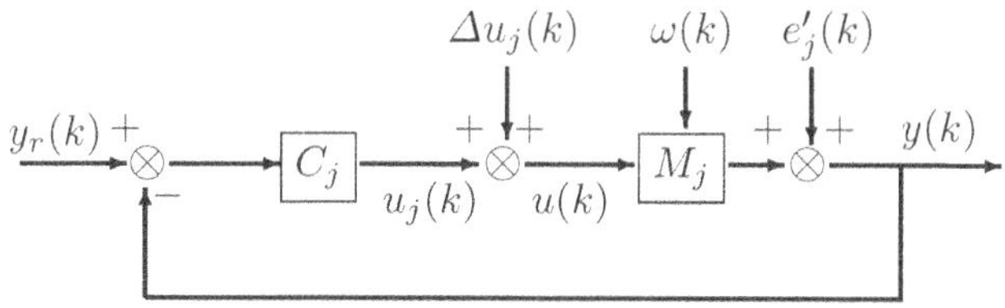

Fig. 4.3 Virtual equivalent system II of WMMAC system

M_j is not the true model of the plant, but the closest one to the true model of plant P.

Then we can construct the VES of WMMAC based on C_j and M_j, as shown in Fig. 4.3, where $u(k)$, $\Delta u_j(k)$, $\phi_c^T(k)$, and $\phi^T(k-d)$ are the same as defined in Sect. 4.3.1 for VES I; and $e_j'(k)$ is defined as follows

$$e_j'(k) = y(k) - \phi^T(k-d)\theta_j - \omega(k) = e_j(k) - \omega(k) \tag{4.40}$$

As we will see later in the next section, the properties of $\Delta u_j(k)$ and $e_j'(k)$ together with the stabilizing and tracking characteristics of 'local' control strategy is the key factor to the stability of the WMMAC system.

4.4 Main Result

Based on two types of VESs, this section gives the stability analysis results of the WMMAC system, in which each 'local' controller may be designed by any stabilizing and tracking strategies.

4.4.1 Stability of VES I

Theorem 4.2 *The proposed WMMAC system is stable if only the following conditions are satisfied:*

(1) The true model of plant, say M_j, is included in the model set $\mathbb{M}$;
(2) Model M_j generates with probability one the minimum output error in the sense that

$$\begin{cases} \sum_{r=1}^{k} e_j^2(r) < \sum_{r=1}^{k} e_i^2(r), \forall k \geq k^*, i \neq j \\ \lim_{k\to\infty} \frac{1}{k} \sum_{r=1}^{k} e_j^2(r) = R_j; \lim_{k\to\infty} \frac{1}{k} \sum_{r=1}^{k} e_i^2(r) = R_i, R_j < R_i, i \neq j \end{cases}$$

where k^* is an unknown limited time instant, R_j is a constant, R_i may be constant or infinity.

(3) *Each 'local' controller is well defined such that C_i is stabilizing M_i, $i = 1, \ldots, j, \ldots N$, and the output of the resulting closed loop system $\{C_i, M_i\}$, say $y_d(k)$, is tracking the reference signal $y_r(k)$ in the sense that*

$$\lim_{k\to\infty} \frac{1}{k} \sum_{i=1}^{k} [y_d(i) - y_r(i)]^2 = R', \; R \le R' < \infty$$

Remark 4.1 R' may achieve its minimum value if the plant is minimum phase and the 'local' controller is designed according to minimum variance principle.

$y_d(k)$ generally refers to the output of each closed-loop system $\{C_i, M_i\}$ which exists only in design.

Proof First, according to Theorem 4.1, Condition (2) guarantees

$$\lim_{k\to\infty} p_j(k) = 1; \; \lim_{k\to\infty} p_i(k) = 0, i = 1, \ldots N, i \neq j \tag{4.41}$$

where j indexes the true model of plant P .

Then we know that the WMMAC system in this situation is equivalent to VES I in the input-output sense.

Next, VES I as shown in Fig. 4.2 can be decomposed into two subsystems, as shown in Figs. 4.4 and 4.5, respectively.

Due to the fact that VES I is a linear time-invariant (LTI) system in structure, we have

$$y(k) = y_1(k) + y_2(k) \tag{4.42}$$

$$u(k) = u_1(k) + u_2(k) \tag{4.43}$$

where $y_1(k) = 0$, $u_1(k) = 0$, $y_2(k) = 0$, $u_2(k) = 0$ for $k < 0$.

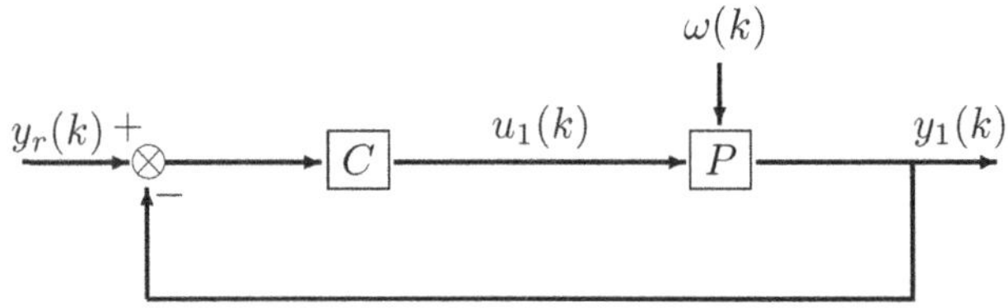

Fig. 4.4 Subsystem 1 of VES I

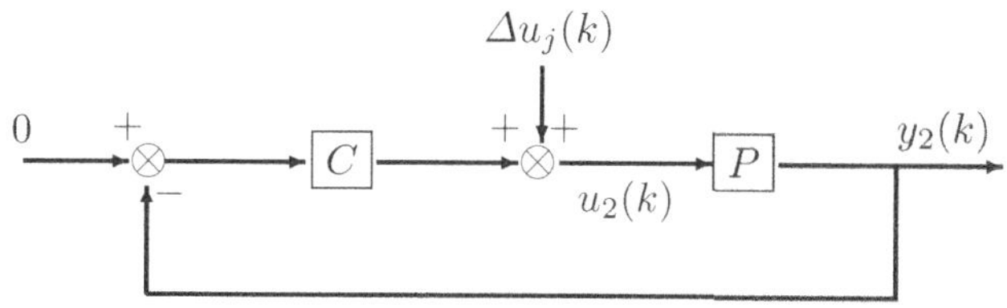

Fig. 4.5 Subsystem 2 of VES I

To facilitate the proof, we need to define a new vector $\widetilde{\phi}(k)$, whose elements are the union of that of $\phi(k-d)$ and $\phi_c(k)$. Without loss of generality, we assume $s_1 < n, s_2 < m$, then $\widetilde{\phi}(k)$ takes the form of

$$\begin{aligned}\widetilde{\phi}(k) = [y(k), \ldots, y(k-n), u(k-1), \ldots, \\ u(k-d-m), y_r(k), \ldots, y_r(k-s_3)]\end{aligned} \tag{4.44}$$

Similarly, define its counterparts $\widetilde{\phi}_1(k)$ in subsystem 1 (Fig. 4.4), and $\widetilde{\phi}_2(k)$ in subsystem 2 (Fig. 4.5), respectively, i. e.

$$\begin{aligned}\widetilde{\phi}_1(k) = [y_1(k), \ldots, y_1(k-n), u_1(k-1), \ldots, \\ u_1(k-d-m), y_r(k), \ldots, y_r(k-s_3)]\end{aligned} \tag{4.45}$$

$$\begin{aligned}\widetilde{\phi}_2(k) = [y_2(k), \ldots, y_2(k-n), u_2(k-1), \ldots, \\ u_2(k-d-m), 0, \ldots, 0]\end{aligned} \tag{4.46}$$

Then, we have

$$\widetilde{\phi}(k) = \widetilde{\phi}_1(k) + \widetilde{\phi}_2(k) \tag{4.47}$$

$$\|\phi(k-d)\| = O(\|\widetilde{\phi}(k)\|) \tag{4.48}$$

$$\|\phi_c(k)\| = O(\|\widetilde{\phi}(k)\|) \tag{4.49}$$

The Big-Oh operator is defined in Appendix.

Obviously, subsystem 1 as shown in Fig. 4.4 is a time-invariant stochastic system, and Condition (3) guarantees that the closed-loop system is stable and tracking. That means

$$\lim_{k\to\infty} \frac{1}{k} \sum_{i=1}^{k} \|\widetilde{\phi}_1(i)\|^2 < \infty \tag{4.50}$$

$$\lim_{k\to\infty} \frac{1}{k} \sum_{i=1}^{k} [y_1(i) - y_r(i)]^2 = R' \tag{4.51}$$

Subsystem 2 as shown in Fig. 4.5 is a stable deterministic system with input signal given by (4.38). Considering (4.49), we obtain by Theorem 14 on p. 111 of reference [22]

$$|y_2(k)| = O(|\Delta u_j(k)|) = o(\|\phi_c(k)\|) = o(\|\widetilde{\phi}(k)\|) \tag{4.52}$$

$$|u_2(k)| = O(|\Delta u_j(k)|) = o(\|\phi_c(k)\|) = o(\|\widetilde{\phi}(k)\|) \tag{4.53}$$

Further, we have

$$\frac{1}{k}\sum_{i=1}^{k}[y_2(i)]^2 = o\left(\frac{1}{k}\sum_{i=1}^{k}\|\widetilde{\phi}(i)\|^2\right) \tag{4.54}$$

$$\frac{1}{k}\sum_{i=1}^{k}[u_2(i)]^2 = o\left(\frac{1}{k}\sum_{i=1}^{k}\|\widetilde{\phi}(i)\|^2\right) \tag{4.55}$$

Equations (4.54) and (4.55) imply

$$\frac{1}{k}\sum_{i=1}^{k}\|\widetilde{\phi}_2(i)\|^2 = o\left(\frac{1}{k}\sum_{i=1}^{k}\|\widetilde{\phi}(i)\|^2\right) \tag{4.56}$$

Then we conclude by Lemma A.2 in Appendix A

$$\lim_{k\to\infty}\frac{1}{k}\sum_{i=1}^{k}\|\widetilde{\phi}(i)\|^2 < \infty \tag{4.57}$$

That means the boundedness of the input-output signals of the WMMAC system.

Next we turn to the tracking performance of the WMMAC system.

Considering (4.54) and (4.57), it is obvious that

$$\frac{1}{k}\sum_{i=1}^{k}[y_2(i)]^2 = o(1) \tag{4.58}$$

Further, by Lemma A.4 in Appendix A, we have

$$\begin{aligned}\lim_{k\to\infty}\frac{1}{k}\sum_{i=1}^{k}[y(i)-y_r(i)]^2 &= \lim_{k\to\infty}\frac{1}{k}\sum_{i=1}^{k}[y_1(i)-y_r(i)+y_2(i)]^2 \\ &= \lim_{k\to\infty}\frac{1}{k}\sum_{i=1}^{k}[y_1(i)-y_r(i)]^2\end{aligned} \tag{4.59}$$

i.e.,

$$\lim_{k\to\infty}\frac{1}{k}\sum_{i=1}^{k}[y(i)-y_r(i)]^2 = R' \tag{4.60}$$

That completes the proof of Theorem 4.2.

4.4.2 Stability of VES II

Theorem 4.3 *The proposed WMMAC system is stable if only the following conditions are satisfied:*

(1) $M_j \in \mathbb{M}$ *is the model closest to the true plant in the following sense with probability one*

$$\begin{cases}\sum_{r=1}^{k} e_j^2(r) < \sum_{r=1}^{k} e_i^2(r), \forall k \geq k^*, i \neq j \\ \lim_{k\to\infty}\frac{1}{k}\sum_{r=1}^{k} e_j^2(r) = R_j;\ \lim_{k\to\infty}\frac{1}{k}\sum_{r=1}^{k} e_i^2(r) = R_i,\ R_j < R_i, i \neq j\end{cases}$$

where k^* *is an unknown limited time instant,* R_j *is a constant,* R_i *may be constant or infinity.*

(2) Each 'local' controller is well defined such that C_i *is stabilizing* M_i, $i = 1, \ldots, j, \ldots N$, *and the output of the resulting closed loop system* $\{C_i, M_i\}$, *say* $y_d(k)$, *is tracking the reference signal* $y_r(k)$ *in the sense that*

$$\lim_{k\to\infty}\frac{1}{k}\sum_{i=1}^{k}[y_d(i)-y_r(i)]^2 = R',\ R \leq R' < \infty$$

(3) For the closest model M_j, *we have*

$$|e'_j(k)| = |e_j(k) - \omega(k)| = o(\|\phi(k-d)\|)$$

Proof First, according to Theorem 4.1, Condition (1) guarantees

$$\lim_{k\to\infty} p_j(k) = 1;\ \lim_{k\to\infty} p_i(k) = 0, i = 1, \ldots N, i \neq j \tag{4.61}$$

where j indicates the closest model to the true plant P, i.e., M_j.

Then, the WMMAC system in this situation is equivalent to VES II in the input-output sense.

Next, we decompose VES II (Fig. 4.3) into three subsystems, as shown in Figs. 4.6, 4.7, and 4.8, respectively.

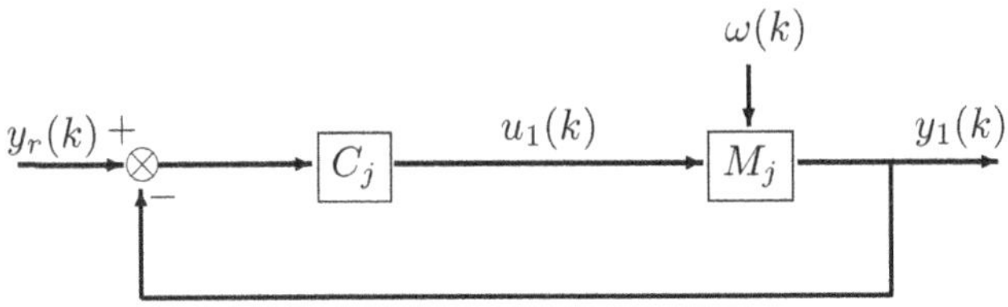

Fig. 4.6 Subsystem 1 of VES II

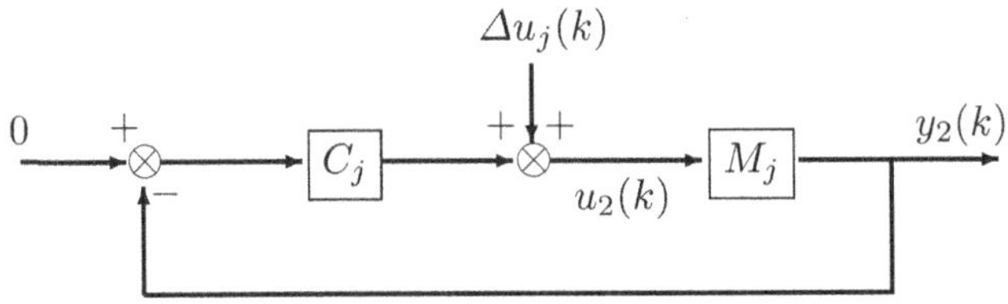

Fig. 4.7 Subsystem 2 of VES II

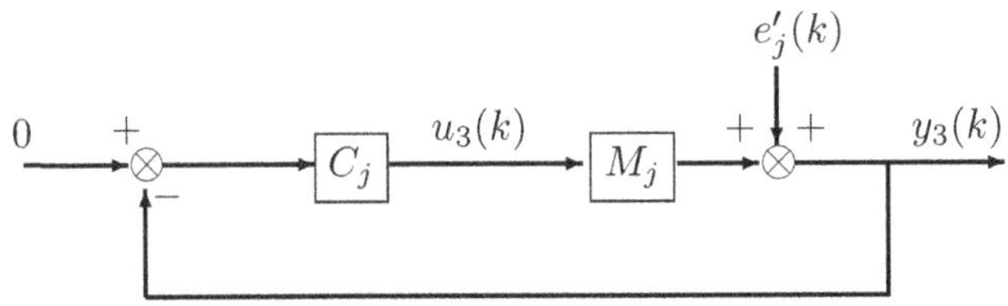

Fig. 4.8 Subsystem 3 of VES II

By linear superposition principle we get

$$y(k) = y_1(k) + y_2(k) + y_3(k) \tag{4.62}$$

$$u(k) = u_1(k) + u_2(k) + u_3(k) \tag{4.63}$$

Similar to the proof of Theorem 4.2, we define $\widetilde{\phi}(k)$, $\widetilde{\phi}_1(k)$, $\widetilde{\phi}_2(k)$, and $\widetilde{\phi}_3(k)$. In detail, $\widetilde{\phi}(k)$, $\widetilde{\phi}_1(k)$, $\widetilde{\phi}_2(k)$ are the same as (4.44), (4.45), (4.46), respectively, and

$$\begin{aligned}\widetilde{\phi}_3(k) = [&y_3(k), \ldots, y_3(k-n), u_3(k-1), \ldots, \\ &u_3(k-d-m), 0, \ldots, 0]\end{aligned} \tag{4.64}$$

Then we obtain

$$\widetilde{\phi}(k) = \widetilde{\phi}_1(k) + \widetilde{\phi}_2(k) + \widetilde{\phi}_3(k) \tag{4.65}$$

$$\|\phi_c(k)\| = O(\|\widetilde{\phi}(k)\|) \tag{4.66}$$

$$\|\phi(k-d)\| = O(\|\widetilde{\phi}(k)\|) \tag{4.67}$$

First, by Condition (2), we know that subsystem 1 (Fig. 4.6) is a stable LTI stochastic system, that means

$$\lim_{k\to\infty}\frac{1}{k}\sum_{i=1}^{k}\|\widetilde{\phi}_1(i)\|^2<\infty \tag{4.68}$$

$$\lim_{k\to\infty}\frac{1}{k}\sum_{i=1}^{k}[y_1(i)-y_r(i)]^2=R' \tag{4.69}$$

Second, subsystem 2 (Fig. 4.7) is a stable deterministic system with input signal given by (4.38).

Thus, we have

$$|y_2(k)|=O(|\Delta u_j(k)|)=o(\|\phi_c(k)\|)=o(\|\widetilde{\phi}(k)\|) \tag{4.70}$$

$$|u_2(k)|=O(|\Delta u_j(k)|)=o(\|\phi_c(k)\|)=o(\|\widetilde{\phi}(k)\|) \tag{4.71}$$

Further we get

$$\frac{1}{k}\sum_{i=1}^{k}[y_2(i)]^2=o\left(\frac{1}{k}\sum_{i=1}^{k}\|\widetilde{\phi}(i)\|^2\right) \tag{4.72}$$

$$\frac{1}{k}\sum_{i=1}^{k}[u_2(i)]^2=o\left(\frac{1}{k}\sum_{i=1}^{k}\|\widetilde{\phi}(i)\|^2\right) \tag{4.73}$$

$$\frac{1}{k}\sum_{i=1}^{k}\|\widetilde{\phi}_2(i)\|^2=o\left(\frac{1}{k}\sum_{i=1}^{k}\|\widetilde{\phi}(i)\|^2\right) \tag{4.74}$$

Finally, let's consider subsystem 3 (Fig. 4.8), which is also a stable deterministic system.

According to Condition (3), the input signal of subsystem 3 has the following property

$$\begin{aligned}\frac{1}{k}\sum_{i=1}^{k}[e_j'(i)]^2&=o\left(\frac{1}{k}\sum_{i=1}^{k}\|\phi(i-d)\|^2\right)\\&=o\left(\frac{1}{k}\sum_{i=1}^{k}\|\widetilde{\phi}(i)\|^2\right)\end{aligned} \tag{4.75}$$

By the fact that subsystem 3 is stable, we obtain

$$|y_3(k)|=O(|e_j'(k)|) \tag{4.76}$$

$$\|u_3(k)\| = O(|e'_j(k)|) \tag{4.77}$$

Further we have

$$\frac{1}{k}\sum_{i=1}^{k}[y_3(i)]^2 = o\left(\frac{1}{k}\sum_{i=1}^{k}\|\widetilde{\phi}(i)\|^2\right) \tag{4.78}$$

$$\frac{1}{k}\sum_{i=1}^{k}[u_3(i)]^2 = o\left(\frac{1}{k}\sum_{i=1}^{k}\|\widetilde{\phi}(i)\|^2\right) \tag{4.79}$$

$$\frac{1}{k}\sum_{i=1}^{k}\|\widetilde{\phi}_3(i)\|^2 = o\left(\frac{1}{k}\sum_{i=1}^{k}\|\widetilde{\phi}(i)\|^2\right) \tag{4.80}$$

Then by Lemma A.2 in Appendix A, regarding $\widetilde{\phi}_2(k)$ and $\widetilde{\phi}_3(k)$ as one variate, we obtain

$$\lim_{k\to\infty}\frac{1}{k}\sum_{i=1}^{k}\|\widetilde{\phi}(i)\|^2 < \infty \tag{4.81}$$

That means the boundedness of the input-output signals of the WMMAC system.

Further by Lemma A.4 in Appendix A, we have the tracking performance of VES II, i.e.

$$\begin{aligned}\lim_{k\to\infty}\frac{1}{k}\sum_{i=1}^{k}[y(i)-y_r(i)]^2 &= \lim_{k\to\infty}\frac{1}{k}\sum_{i=1}^{k}[y_1(i)-y_r(i)]^2\\ &= R'\end{aligned} \tag{4.82}$$

That completes the proof of Theorem 4.3.

Similar to the proof of Theorem 4.3 (so the details are omitted), we have the following corollary for a general WMMAC system.

Corollary 4.1 *The proposed WMMAC system is stable if only the following conditions are satisfied:*

(1) $M_j \in \mathbb{M}$ *is the model closest to the true plant in the following sense with probability one*

$$\begin{cases}\sum_{r=1}^{k} e_j^2(r) < \sum_{r=1}^{k} e_i^2(r), \forall k \geq k^*, i \neq j\\ \lim_{k\to\infty}\frac{1}{k}\sum_{r=1}^{k} e_j^2(r) = R_j;\ \lim_{k\to\infty}\frac{1}{k}\sum_{r=1}^{k} e_i^2(r) = R_i, R_j < R_i, i \neq j\end{cases}$$

where k^* *is an unknown limited time instant,* R_j *is a constant,* R_i *may be constant or infinity.*

(2) Each 'local' controller is well defined such that C_i *is stabilizing its corresponding local model* M_i, $i = 1, \ldots, j, \ldots N$, *and the output of the resulting closed-loop system* $\{C_i, M_i\}$, *say* $y_d(k)$, *is tracking the reference signal* $y_r(k)$

in the sense that

$$\lim_{k\to\infty}\frac{1}{k}\sum_{i=1}^{k}[y_d(i)-y_r(i)]^2 = R', R \le R' < \infty$$

(3) For the closest model M_j, we have

$$\frac{1}{k}\sum_{i=1}^{k}[e'_j(i)]^2 = \frac{1}{k}\sum_{i=1}^{k}[e_j(i)-\omega(i)]^2 = o\left(\frac{1}{k}\sum_{i=1}^{k}\|\phi(i-d)\|^2\right)$$

Remark 4.2 Although we only considered SISO system, it is straightforward to develop the same results for multi-input multi-output (MIMO) system, because we adopted norm operation to draw the theorems and the corollary.

4.5 Simulation Results

Consider a discrete-time plant with parametric uncertainties

$$(1 + a_1q^{-1} + a_2q^{-2})y(k) = q^{-1}(b_0 + b_1q^{-1})u(k) + \omega(k) \tag{4.83}$$

where $\omega(k)$ is a zero-mean white noise sequence that was created with the **Matlab randn** function. The deterministic part of (4.83) is obtained by converting the following continuous-time LTI model to a discrete-time model with sample time $t_s = 0.5s$ and the zero order hold.

$$\frac{k}{s^2 - 3s + 2} \tag{4.84}$$

The uncertainties of (4.83) originates from k in (4.84). For simplicity, we suppose there are only four possible situations as the uncertainties of k, i.e., $k = 0.7, k = 0.8$, $k = 1$, and $k = 0.9$. That means in (4.83), a_1 and a_2 are constants, i.e., $a_1 = -4.3670$, $a_2 = 4.4817$, b_0 and b_1 depend on k. Four 'local' controllers were designed by pole assignment strategy. Each controller stabilizes a possible model by formulating an expected closed-loop characteristic polynomial, say $A_m(q^{-1})$ and track the reference signal $y_r(k)$.

In detail, local controller 1 is designed according to local model 1 ($k = 0.7$), i.e.

$$(1 - 4.3670q^{-1} + 4.4817q^{-2})y(k) = q^{-1}(0.1473 + 0.2428q^{-1})u(k) + \omega(k) \tag{4.85}$$

Local controller 2 is designed according to local model 2 ($k = 0.8$), i.e.

$$(1 - 4.3670q^{-1} + 4.4817q^{-2})y(k) = q^{-1}(0.1683 + 0.2775q^{-1})u(k) + \omega(k) \tag{4.86}$$

Local controller 3 is designed according to local model 3 ($k = 1$), i.e.

$$(1 - 4.3670q^{-1} + 4.4817q^{-2})y(k) = q^{-1}(0.2104 + 0.3469q^{-1})u(k) + \omega(k) \tag{4.87}$$

Local controller 4 is designed according to local model 4 ($k = 0.9$), i.e.

$$(1 - 4.3670q^{-1} + 4.4817q^{-2})y(k) = q^{-1}(0.1894 + 0.3122q^{-1})u(k) + \omega(k) \tag{4.88}$$

The expected closed-loop characteristic polynomial is chosen to be

$$A_m(q^{-1}) = 1 - 1.3205q^{-1} + 0.4966q^{-2} \tag{4.89}$$

which corresponds to the characteristic polynomial of the following continuous-time second-order system

$$\frac{{\omega_n}^2}{s^2 + 2\xi\omega_n s + {\omega_n}^2} \tag{4.90}$$

with $\xi = 0.707$, $\omega_n = 1$, and sample time $t_s = 0.5$ s.

Case 1: The true model of plant is included in the model set, say Model 2 ($k = 0.8$); the variance of $\omega(k)$ is chosen to be $\sigma = 0.1$. The simulation results, i.e., the four weights signals, the closed-loop output $y(k)$ against reference signal $y_r(k)$, and the control signal $u(k)$ are shown in Figs. 4.9 and 4.10.

Case 2: The true model of plant is not included in the model set, which corresponds to $k = 1.03$ in (4.84), i.e.

$$(1 - 4.3670q^{-1} + 4.4817q^{-2})y(k) = q^{-1}(0.2167 + 0.3573q^{-1})u(k) + \omega(k) \tag{4.91}$$

Model 3 is the closest in the model set to model (4.91); the variance of $\omega(k)$ is chosen to be $\sigma = 0.1$. The simulation results are shown in Figs. 4.11 and 4.12.

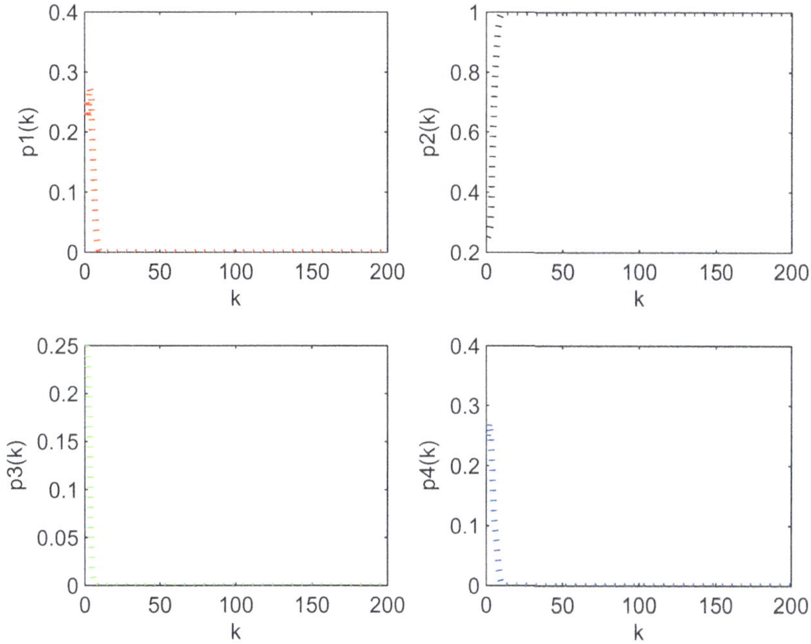

Fig. 4.9 Controller weight signals of Case 1

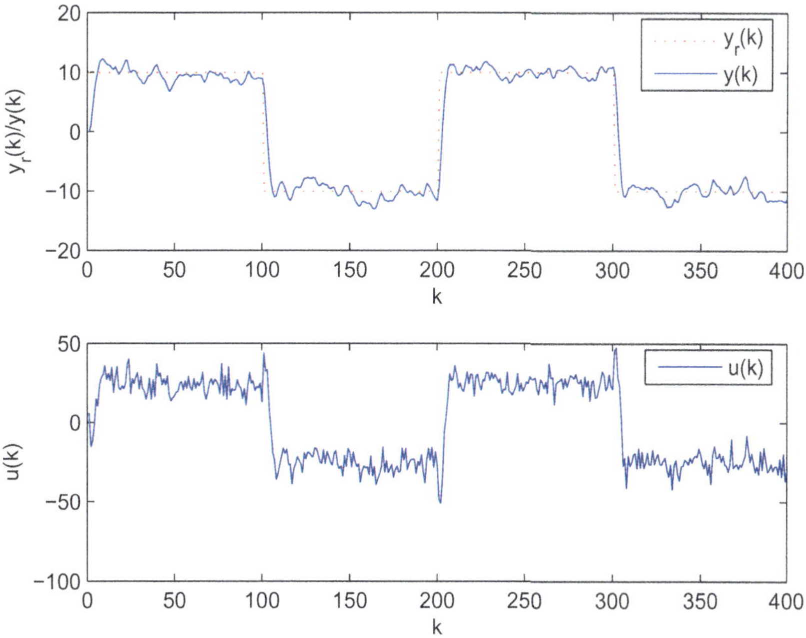

Fig. 4.10 Output, reference, and control signals of Case 1

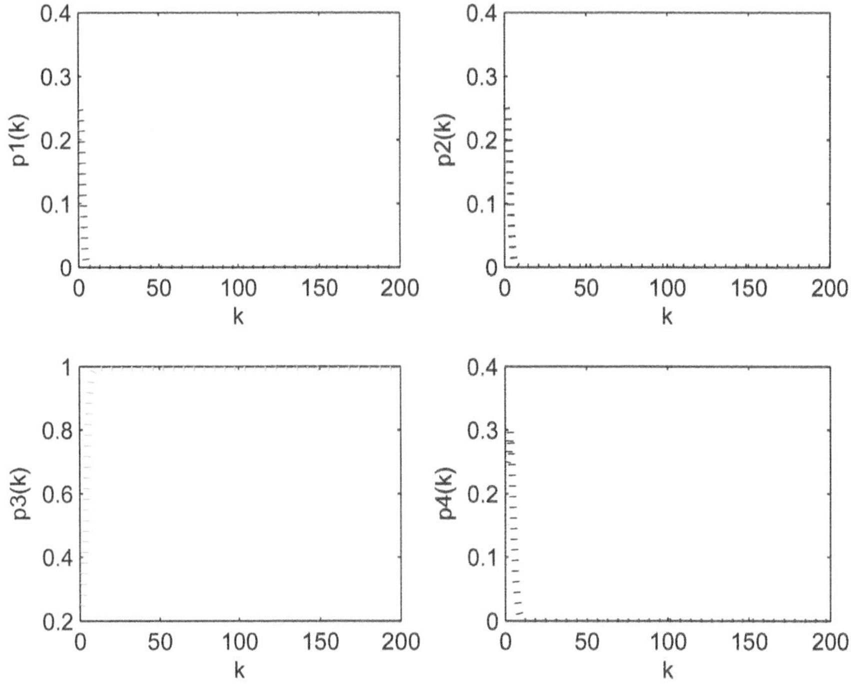

Fig. 4.11 Controller weight signals of Case 2

In summary, the four weights $p_i(k)$, $i = 1, 2, 3, 4$ in the proposed weighting algorithm converge correctly in each case, and the closed-loop WMMAC system is stable.

However, as shown in Case 3, if the noise level is high enough, while the difference between the true model of plant and each of the other models is not significant, then the WMMAC system falls into 'model-identification confusion', i.e., weighting algorithm can not converge correctly. Consequently, the WMMAC system performance will be drastically degraded, since related theoretical assumptions, i.e., (4.14) and (4.15), were severely violated.

Case 3: Suppose there are four possible situations as the uncertainties of k in (4.84), i.e., $k = 0.97$, $k = 0.98$, $k = 1$, and $k = 0.99$. The true model of plant is not included in the model set, which corresponds to $k = 1.01$ in (4.84). The variance of $\omega(k)$ is $\sigma = 10$. The corresponding simulation results are demonstrated in Figs. 4.13, and 4.14 respectively.

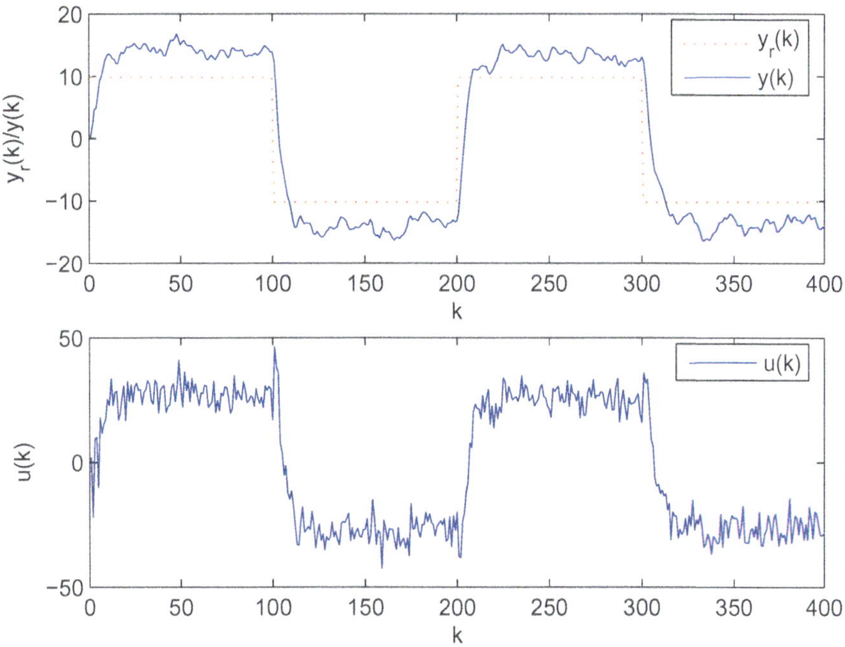

Fig. 4.12 Output, reference, and control signals of Case 2

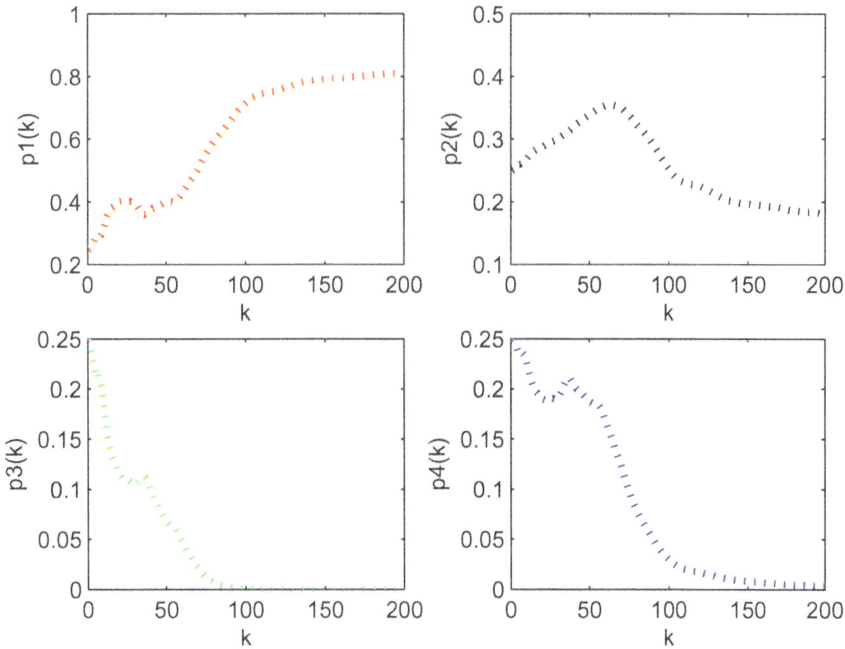

Fig. 4.13 Controller weight signals of Case 3

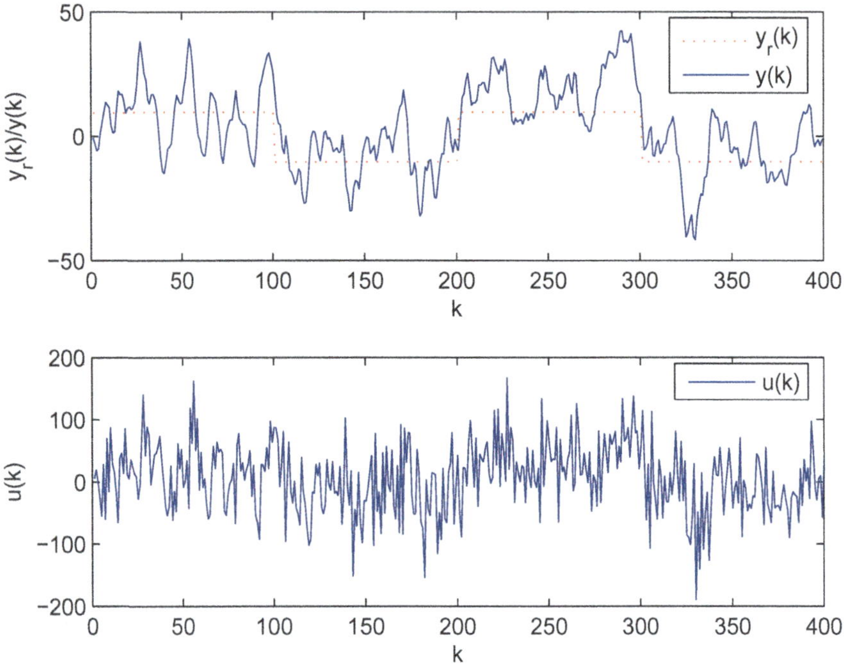

Fig. 4.14 Output, reference, and control signals of Case 3

4.6 Conclusion

Owing to a new weighting algorithm and a new analysis tool, i.e., VES approach, the long-standing stability issue of WMMAC (including RMMAC) systems, has been addressed under some smooth conditions. To some extent, the corresponding stability criteria on WMMAC systems are independent of specific 'local' controller strategy and weighting algorithm. Thus, in the author's opinion, VES concept and methodology could be a unified theoretical framework for understanding and evaluating different kinds of WMMAC (including RMMAC) schemes. Simulation results verified the effectiveness of the proposed scheme, including weighting algorithm and stability analysis results. Of course, there are still some room for improvement, for example, the conditions for convergence of weighting algorithm, and the convergence rate of weighting algorithm.

References

1. D.T. Magill, Optimal adaptive estimation of sampled stochastic processes. IEEE Trans. Autom. Control **10**, 434–439 (1965)
2. D.G. Lainiotis, Partitioning: aunifying framework for adaptive systems I: estimation; II: control, in *Proceedings of IEEE*, vol. 64 (1976), pp. 1126–1143 and 1182–1197
3. M. Athans et al., The stochastic control of the F-8C aircraft using a multiple model adaptive control (MMAC) method-Part I: equilibrium flight. IEEE Trans. Autom. Control **22**, 768–780 (1977)
4. D.W. Lane, P.S. Maybeck, Multiple model adaptive estimation applied to the Lambda URV for failure detection and identification, in *Proceedings of 33rd IEEE Conference on Decision and Control* (Lake Buena Vista, FL, 1994), pp. 678–683
5. C. Yu, R.J. Roy, H. Kaufman, B.W. Bequette, Multiple-model adaptive predictive control of mean arterial pressure and cardiac output. IEEE Trans. Biomed. Eng. **39**, 765–778 (1992)
6. R.L. Moose, H.F. Van Landingham, D.H. McCabe, Modeling and estimation for tracking maneuvering targets. IEEE Trans. Aerospace Elec. Syst. **AES-15**, 448–456 (1979)
7. X.R. Li, Y. Bar-Shalom, Design of an interacting multiple model algorithm for air traffic control tracking. IEEE Trans. Contr. Syst. Tech. **1**, 186–194 (1993)
8. M. Athans, S. Fekri, A. Pascoal, Issues on robust adaptive feedback control, in *Preprints 16th IFAC World Congress, Invited Plenary paper* (Prague, Czech Republic, 2005), pp. 9–39
9. S. Fekri, M. Athans, A. Pascoal, Issues, progress and new results in robust adaptive control. Int. J. Adapt. Control Signal Process **20**(10), 519–579 (2006)
10. S. Fekri, M. Athans, A. Pascoal, Robust multiple model adaptive control (RMMAC): a case study. Int. J. Adapt. Control Signal Process **21**(1), 1–30 (2007)
11. Y. Baram, Information, consistent estimation and dynamic system identification. Ph.D. Dissertation (MIT, Cambridge, MA, USA, 1976)
12. Y. Baram, N.R. Sandell, An information theoretic approach to dynamical systems modeling and identification. IEEE Trans. Autom. Control **23**(1), 61–66 (1978)
13. Y. Baram, N.R. Sandell, Consistent estimation on finite parameter sets with application to linear systems identification. IEEE Trans. Autom. Control **23**(3), 451–454 (1978)
14. A. Kehagias, Convergence properties of the lainiotis partition algorithm. Control Comput. **19**, 1–6 (1991)
15. P. Ioannou, Robust adaptive control: the search for the Holy Grail, in *Proceedings of 47th IEEE Conference on Decision and Control* (Cancun, Mexico, 2008), pp. 12–13
16. V. Hassani, J. Hespanha, M. Athans, A. Pascoal, Stability analysis of robust multiple model adaptive control, in *Proceedings of 18th IFAC World Congress* (Milan, Italy, 2011)
17. W. Zhang, Theoretical research and application of robust adaptive control. Ph. D Thesis (Tsinghua University, 1993)
18. W. Zhang, T.G. Chu, L. Wang, A new theoretical framework for self-tuning control. Int. J. Inf. Technol. **11**(11), 123–139 (2005)
19. W. Zhang, X.L. Li, J.Y. Choi, A unified analysis of switching multiple model adaptive control—Virtual equivalent system approach, in *Proceedings of 17th IFAC World Congress* (Seoul, Korea, 2008)
20. W. Zhang, On the stability and convergence of self-tuning control-virtual equivalent system approach. Int. J. Control **83**(5), 879–896 (2010)
21. K.S. Narendra, Adaptive control of discrete-time systems using multiple models. IEEE Trans. Autom. Control **45**(9), 1669–1686 (2000)
22. C.A. Desoer, M. Vidyasagar, *Feedback Systems: Input–Output Properties* (Academic Press, New York, 1975)

Chapter 5
Further Results on Stable Weighted Multiple Model Adaptive Control of Discrete-Time Stochastic Plant

Abstract This chapter is intended to further improve the weighting algorithm and the transient performance of WMMAC of discrete-time stochastic plant. In order to relax the convergence conditions and to further improve the convergence rate of weighting algorithm proposed in Chap. 4, an improved weighting algorithm is proposed in this chapter. The stability and convergence of the corresponding WMMAC systems for two types of stochastic plants are proved according to VES concept and methodology. The first type of stochastic plant is linear time-invariant system with unknown parameters, the second is linear time-varying system with jumping parameters. Finally, some simulation results are presented to verify the effectiveness of theoretical results and the satisfactory performance of the closed-loop WMMAC system.

5.1 Introduction

As we know, weighting algorithm plays very important role in the WMMAC system, especially in transient performance. In fact, according to the analysis results of Chap. 4 (also see reference [1]), the closed-loop stability of a WMMAC system mainly depends on three conditions:

(1) the model set is constructed correctively, i.e., it includes the true model of the plant or a closet one to the plant;
(2) the weighting algorithm converges correctly to identify the true model of the plant or the closest model to the plant;
(3) each local controller is constructed to stabilize its corresponding local model and tracking the reference signal.

In addition, simulation results shows that the convergence rate of the weighting algorithm will have effect on the transient performance of the corresponding WMMAC system.

W. Zhang and Q. Li, *Virtual Equivalent System Approach for Stability Analysis of Model-based Control Systems*, https://doi.org/10.1007/978-981-15-5538-1_5

In classical WMMAC/WMMAE systems [1–8], including RMMAC systems [9–11], probabilistic weighting algorithm was widely adopted, which is based on a bank of Kalman filters, dynamic hypothesis testing, and Bayes' formula. Some convergence results on the probabilistic weighting algorithm have been obtained in early stage of multiple model research [12–15]. In detail, Fekri, Athans and Pascoal pointed out in [10] that under certain ergodicity and stationarity assumptions on model output errors, one of the posterior probabilities will converge almost surely to unity and will 'identify' the model closest to the true plant, i.e. the one with smallest Baram Proximity Measure (BPM) [10, 12].

However, due to using Kalman filters to drive the weighting algorithm, the classical WMMAC scheme may suffer from poor performance due to either large initial state estimate error or inaccurate knowledge of the disturbance/noise statistics. Additionally, the complexity and calculation burden of the supervisor may hinder its application because every candidate controller requires a Kalman filter and a post posterior evaluation [16]. Thus, some substitutive methods have been proposed to improve the above-mentioned situation. In [17, 18], fuzzy rules based weighting algorithms replaced the probabilistic weighting algorithm. In [1], a model output error based weighting algorithm is proposed, and furthermore, the long-standing stability issue of WMMAC has been proved for a class of discrete-time stochastic plant under some certain conditions.

This chapter is aimed to further relax the convergence requirements and to further improve the convergence rate of weighting algorithm proposed in Chap. 4 and reference [1]. Besides, in contrast to Chap. 4 and reference [1], more general stability and convergence results of WMMAC systems are also presented for linear stochastic plants with unknown time-invariant parameters or with jumping parameters.

5.2 Description of a WMMAC System

Although there are different types of WMMAC systems, we use, for analysis purpose, a concise block diagram as shown in Fig. 5.1 to represent a general WMMAC system of discrete-time stochastic plant, in which the details of 'local' control strategy and weighting algorithm are omitted for simplicity.

In Fig. 5.1, C_i represents a local controller; $p_i(k)$ represents the weighting coefficient corresponding to the ith plant model M_i and then acts as the controller weight of C_i. Denote $C_i \in \mathbb{C}$. $\mathbb{C} = \{C_i, i = 1, 2, \ldots, N\}$ is the controller set. And C_i is designed according to model $M_i \in \mathbb{M}$. $\mathbb{M} = \{M_i, i = 1, 2, \ldots, N\}$ is the model set that covers the uncertainties of plant P. The reference input $y_r(k)$ is bounded, i.e., $y_r(k) < \infty$.

As to the local control strategy, any existing design methods are adoptable, such as pole assignment (placement), mixed-μ synthesis tools, etc. Each 'local' controller

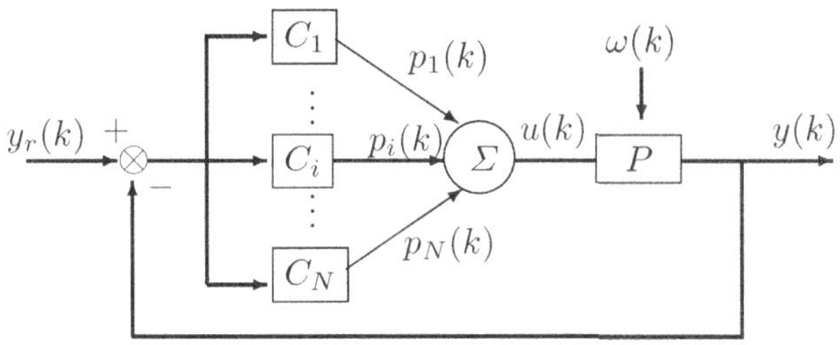

Fig. 5.1 Simplified block diagram of a general WMMAC system

C_i generates local control signal $u_i(k)$; the 'global' control signal applied to the plant is obtained by the weighting sum of all the local control signals, i.e.

$$u(k) = \sum_{i=1}^{N} p_i(k)u_i(k) \tag{5.1}$$

where $p_i(k), i = 1, 2, \ldots, N$ represents ith controller weight.

A simple and direct weighting algorithm is proposed in reference [1], which depends only on each local model output error. The new weighting algorithm gets rid of the inconvenience of Kalman filters as well as off-line membership functions.

For simplicity, we first consider the following discrete-time stochastic plant with single input and single output (SISO)

$$P : A(q^{-1})y(k) = q^{-d}B(q^{-1})u(k) + \omega(k) \tag{5.2}$$

where

$$A(q^{-1}) = 1 + a_1q^{-1} + \cdots + a_{na}q^{-na}$$
$$B(q^{-1}) = b_0 + b_1q^{-1} + \cdots + b_{nb}q^{-nb}$$

$$d \geq 1; na \geq 1; nb \geq 1$$

where $y(k)$, $u(k)$ and $\omega(k)$ are the output, input and zero-mean white noise of the system, respectively, $y(k) = 0, u(k) = 0, \omega(k) = 0$ for $k < 0$, and that

$$\lim_{n\to\infty} \frac{1}{n} \sum_{i=1}^{n} [\omega(i)]^2 = R < \infty \tag{5.3}$$

The plant P can be stable or non-stable, minimum phase or non-minimum phase. Its output $y(k)$ can be rewritten as

$$y(k) = \phi^T(k-d)\theta + \omega(k) \tag{5.4}$$

where

$$\phi^T(k-d) = [y(k-1), \ldots, y(k-na), u(k-d), \ldots, u(k-d-nb)] \tag{5.5}$$

$$\theta = [-a_1, \ldots, -a_{na}, b_0, b_1, \ldots, b_{nb}] \tag{5.6}$$

For each model $M_i \in \mathbb{M}$, its output is given by

$$y_i(k) = \phi^T(k-d)\theta_i \tag{5.7}$$

where θ_i is the parameter vector of model M_i. Further, define the output error of each model M_i, i.e.

$$e_i(k) = y(k) - y_i(k) = y(k) - \phi^T(k-d)\theta_i \tag{5.8}$$

As we will see later in the next section, $e_i(k)$ is used to calculate $p_i(k)$.

5.3 Weighting Algorithms with Convergence Analysis

First, we review the weighting algorithm with convergence result put forwarded in reference [1]. Then, an improved weighting algorithm will be proposed in order to improve the convergence rate and to relax the convergence conditions.

In addition, it is worth pointing out that all the limit operations in this chapter are in the sense of probability one.

Weighting Algorithm 1 [1]:

$$l_i(0) = \frac{1}{N};\ p_i(0) = l_i(0) \tag{5.9}$$

$$l_i'(k) = 1 + \frac{1}{k}\sum_{p=1}^{k} e_i(p)^2 \tag{5.10}$$

$$l_{min}(k) = \min_i l_i'(k) \tag{5.11}$$

$$l_i(k) = l_i(k-1)\frac{l_{min}(k)}{l_i'(k)} \tag{5.12}$$

$$p_i(k) = \frac{l_i(k)}{\sum_{i=1}^{N} l_i(k)} \tag{5.13}$$

Theorem 5.1 *If $M_j \in \mathbb{M}$ is the closest model in the model set to the true plant in the following sense with probability one*

$$\begin{cases} \sum_{r=1}^{k} e_j^2(r) < \sum_{r=1}^{k} e_i^2(r), \forall k \geq k^*, i \neq j \\ \frac{1}{k}\sum_{r=1}^{k} e_j^2(r) \to R_j; \frac{1}{k}\sum_{r=1}^{k} e_i^2(r) \to R_i, R_j < R_i, i \neq j \end{cases} \tag{5.14}$$

where k^* is an unknown limited time instant, R_j is a constant, R_i may be constant or infinity.

Then the weighting Algorithm 1, i.e., Eqs. (5.9)–(5.13) lead to

$$p_j(k) \to 1; p_i(k) \to 0, i = 1, \ldots N, i \neq j \tag{5.15}$$

It is appropriate to take $k^* = d$, d is the time delay of plant output, as shown in Eq. (5.2).

In order to relax the convergence conditions and to improve the convergence rate of weighting Algorithm 1, a modified weighting algorithm is proposed as follows.

Weighting Algorithm 2:

$$l_i(0) = \frac{1}{N}; p_i(0) = l_i(0) \tag{5.16}$$

$$l_i'(k) = \alpha + \frac{1}{k}\sum_{p=1}^{k} e_i(p)^2 \tag{5.17}$$

$$l_{min}(k) = \min_i l_i'(k) \tag{5.18}$$

$$\beta_i(k) = \frac{l_{min}(k)}{l_i'(k)} \tag{5.19}$$

$$l_i(k) = \begin{cases} l_i(k-1) & \text{if } \beta_i(k) = 1 \\ l_i(k-1)[\beta_i(k)]^{ceil\left(\frac{1}{1-\beta_i(k)}\right)} & \text{if } \beta_i(k) < 1 \end{cases} \tag{5.20}$$

$$p_i(k) = \frac{l_i(k)}{\sum_{i=1}^{N} l_i(k)} \tag{5.21}$$

where $\alpha > 0$ is a small constant to avoid $l_i'(k) = 0$; $ceil(x)$ is the ceiling function that generates the smallest integer not less than x, i.e.,

$$ceil(x) = \min\{n \in \mathbb{Z} | x \leq n\}$$

Next we analyze the convergence of weighting Algorithm 2, i.e., Eqs. (5.16)–(5.21).

Theorem 5.2 *Suppose there is a model, say $M_j \in \mathbb{M}$, which is the closest one to the true plant in the following sense with probability one*

$$\sum_{r=1}^{k} e_j(r)^2 < \sum_{r=1}^{k} e_i(r)^2, \forall k \geq d, i \neq j \tag{5.22}$$

where d is the time delay of plant output, as shown in (5.2).

Then weighting Algorithm 2, i.e., Eqs. (5.16)–(5.21) has the following property

$$p_j(k) \to 1;\ p_i(k) \to 0, i = 1, \ldots N, i \neq j \tag{5.23}$$

Proof It is obvious that weighting Algorithm 2, i.e., Eqs. (5.16)–(5.21) together with Eq. (5.22) guarantee the following results with probability one

$$\begin{cases} l_{min}(k) = l'_j(k) \\ \frac{l_{min}(k)}{l'_j(k)} = 1 & \forall k \geq d, i \neq j \\ \frac{l_{min}(k)}{l'_i(k)} < 1 \end{cases} \tag{5.24}$$

Further, we know that if $\beta_i(k) = \frac{l_{min}(k)}{l'_i(k)} < 1$, and

$$\lim_{k \to \infty} \beta_i(k) = \lim_{k \to \infty} \frac{l_{min}(k)}{l'_i(k)} = 1, i \neq j \tag{5.25}$$

then we have

$$\lim_{k \to \infty} [\beta_i(k)]^{ceil\left(\frac{1}{1-\beta_i(k)}\right)} = \frac{1}{e} < 1 \tag{5.26}$$

Putting Eqs. (5.20), (5.24), and (5.26) together, we get

$$l_j(k) \to l_j(d);\ l_i(k) \to 0, i \neq j \tag{5.27}$$

Consequently, from Eq. (5.21) we have

$$p_j(k) \to 1;\ p_i(k) \to 0, i \neq j \tag{5.28}$$

That completes the proof of Theorem 5.2.

By comparison between the convergence condition, i.e., Eq. (5.22) of weighting Algorithm 2 with the convergence condition of weighting Algorithm 1, i.e.,

Eq. (5.14), we see that the convergence condition of weighting Algorithm 2 is relaxed. Besides, it is obvious that the convergence rates of both weighting algorithms depend on the convergence rates of $l_i(k)$.

Generally speaking, owing to the property of ceiling function and the fact that $\beta_i(k) = \frac{l_{min}(k)}{l_i'(k)} < 1, k \geq d$, we have

$$ceil\left(\frac{1}{1-\beta_i(k)}\right) \geq 2, \quad [\beta_i(k)]^{ceil\left(\frac{1}{1-\beta_i(k)}\right)} < \beta_i(k), k \geq d$$

Then we know that $l_i(k), k \geq d, i \neq j$ in weighting Algorithm 2 is smaller than $l_i(k), k \geq d, i \neq j$ in weighting Algorithm 1. Thus weighting Algorithm 2 has sharper convergence rate than weighting Algorithm 1. This is also illustrated later in the simulation results of this chapter, especially for the situation when local models are 'close' to each other.

Remark 5.1 It should also be pointed out that weighting Algorithm 2 is more complicated than weighting Algorithm 1. In other words, the advantages of Algorithm 2 over weighting Algorithm 1 are at the cost of more computational burden.

Obviously, both weighting algorithms will be invalid to deal with the extreme situation that each local model generates the same error index $l_i'(k), i = 1, \ldots, N$, of course such situation is avoidable at design stage of a WMMAC systems.

5.4 Construction of VES

This section describes two types of VES [19] of WMMAC corresponding to two different situations. For the first situation, we consider that the true model of plant is included in the model set; For the second situation, we consider that the true model of plant is not included in the model set, but there is a closest one in the model set to the true plant.

With the help of VESs, the stability and convergence of the corresponding WMMAC systems can be proved under some smooth conditions, details will be given in next section.

5.4.1 VES I: The True Model of Plant Is Included in the Model Set

Suppose

$$\begin{cases} p_j(k) \to 1;\ p_i(k) \to 0, i = 1, \dots N, i \neq j \\ M_j \text{ is the true model of the plant } P \end{cases} \tag{5.29}$$

Then we can construct the VES of WMMAC based on $C = C_j$ and $P = M_j$, as shown in Fig. 5.2.

In Fig. 5.2, $u_j(k)$ is the output of local controller C_j.

Without loss of generality, denote

$$u_i(k) = \phi_c^T(k)\theta_{ci} \tag{5.30}$$

where

$$\begin{aligned} \phi_c^T(k) = [&y(k), y(k-1), \dots, y(k-s_1), u(k-1), \\ &\dots, u(k-s_2), y_r(k), \dots, y_r(k-s_3)] \end{aligned} \tag{5.31}$$

is the regression vector form of local control signal $u_i(k)$. The numbers of the elements of $\phi_c^T(k)$, i.e., s_1, s_2, s_3, are limited integers and depend on specific local control strategy; θ_{ci} is the parameter vector form of local controller C_i.

Then $\Delta u(k)$ is the controller output difference between global control $u(k)$ and local control $u_j(k)$, i.e.

$$\begin{aligned} \Delta u(k) &= u(k) - u_j(k) \\ &= \sum_{i=1}^{N} p_i(k) \cdot u_i(k) - u_j(k) \\ &= \sum_{i=1, i\neq j}^{N} p_i(k) \cdot u_i(k) + [p_j(k) - 1] \cdot u_j(k) \end{aligned} \tag{5.32}$$

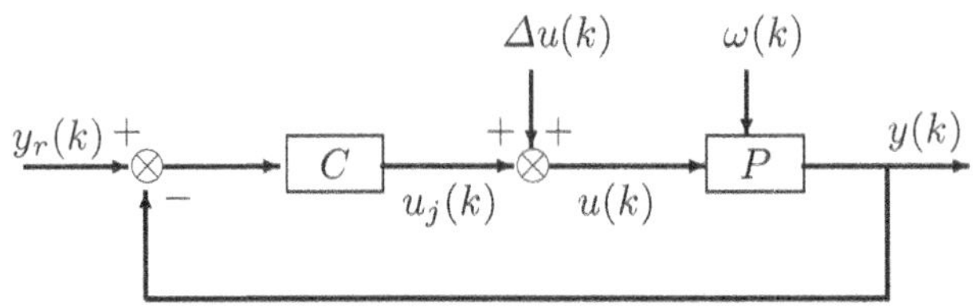

Fig. 5.2 Virtual equivalent system I of a WMMAC system

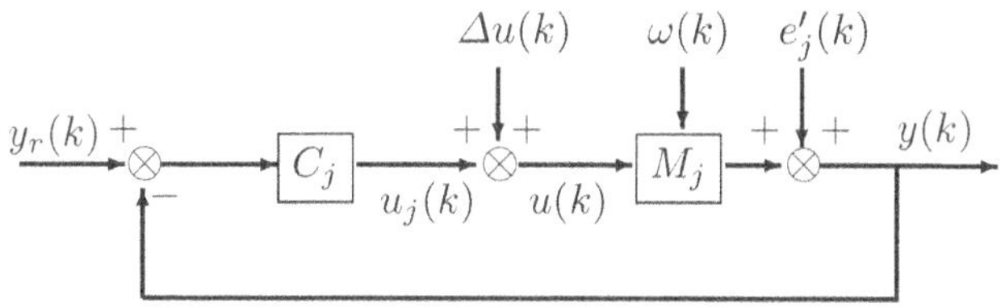

Fig. 5.3 Virtual equivalent system II of a WMMAC system

Based on (5.29), (5.30), and (5.32), it is obvious that

$$\Delta u(k) = o(\|\phi_c(k)\|) \tag{5.33}$$

5.4.2 *VES II: The True Model of Plant Is Not Included in the Model Set*

Suppose

$$\begin{cases} p_j(k) \to 1;\ p_i(k) \to 0,\ i = 1, \ldots N, i \neq j \\ M_j \text{ is the model closest to the true plant } P \end{cases} \tag{5.34}$$

Then we obtain VES II of WMMAC based on C_j and M_j, as shown in Fig. 5.3, where $u(k)$ and $\Delta u(k)$ are the same as in VES I; and $e'_j(k)$ is the output difference between plant P and model M_j, i.e.

$$e'_j(k) = y(k) - \phi^T(k-d)\hat{\theta}_{mj}(k) - \omega(k) = e_j(k) - \omega(k) \tag{5.35}$$

Similar to VES I, we have in Fig. 5.3

$$\|\Delta u(k)\| = o(\|\phi_c(k)\|) \tag{5.36}$$

Next, based on these VESs, we can develop the stability results of the proposed WMMAC system.

5.5 Main Result

This section presents the theoretical analysis results, including three theorems and three corollaries on the stability and convergence of the proposed WMMAC system. These results are given in a general framework. By "general" we mean that the corresponding results are independent of specific local control strategy and specific weighting algorithm.

5.5.1 Stability and Convergence of VES I

Theorem 5.3 *The proposed WMMAC system with weighting Algorithm 2 is stable and convergent if the following conditions are satisfied:*

(1) The true model of plant, M_j, is included in the model set $\mathbb{M}$, which generates the minimum output error in the following sense with probability one

$$\sum_{r=1}^{k} e_j(r)^2 < \sum_{r=1}^{k} e_i(r)^2, \forall k \geq d, i \neq j$$

(2) Each local controller is well defined such that C_i is stabilizing its corresponding local model M_i, $i = 1, \dots, j, \dots N$, and the output of the resulting closed-loop system $\{C_i, M_i\}$, say $y_{out}(k)$, is tracking the reference signal $y_r(k)$ in the sense that

$$\lim_{n\to\infty} \frac{1}{n} \sum_{k=1}^{n} [y_{out}(k) - y_r(k)]^2 = R', R \leq R' < \infty$$

Proof First, Condition (1) guarantees by Theorem 5.2

$$p_j(k) \to 1;\ p_i(k) \to 0, i = 1, \dots N, i \neq j \tag{5.37}$$

where j indexes the true model of plant P.

Then we know that the WMMAC system in this situation is equivalent to VES I in the input-output sense.

Next, it is obvious that VES I as shown in Fig. 5.2 together with the property of $\Delta u(k)$ is totally the same as in the VES I in reference [1].

Then following the similar lines of the proof of Theorem 4.1 in reference [1], we obtain the conclusion of Theorem 5.3.

That completes the proof of Theorem 5.3.

5.5.2 Stability and Convergence of VES II

Theorem 5.4 *The proposed WMMAC system with weighting Algorithm 2 is stable and convergent if the following conditions are satisfied:*

(1) There is a model, say $M_j \in \mathbb{M}$, which is closest to the true plant in the following sense with probability one

$$\sum_{r=1}^{k} e_j(r)^2 < \sum_{r=1}^{k} e_i(r)^2, \forall k \geq d, i \neq j$$

(2) *Each local controller is well defined such that C_i is stabilizing its corresponding local model M_i, $i = 1, \ldots, j, \ldots N$, and the output of the resulting closed-loop system $\{C_i, M_i\}$, say $y_{out}(k)$, is tracking the reference signal $y_r(k)$ in the sense that*

$$\lim_{n\to\infty} \frac{1}{n} \sum_{k=1}^{n} \|y_{out}(k) - y_r(k)\|^2 = R', R \leq R' < \infty$$

(3) *For the model closest to the true plant, i.e., M_j, its output approximates the true plant output in the sense that*

$$\|e'_j(k)\| = \|e_j(k) - \omega(k)\| = o(\|\phi(k-d)\|)$$

Proof First, Condition (1) guarantees by Theorem 5.2

$$p_j(k) \to 1;\ p_i(k) \to 0, i = 1, \ldots N, i \neq j \tag{5.38}$$

where j indicates the model closest to the true plant P, i.e., M_j.

Then we know that the WMMAC system in this situation is equivalent to VES II in the input-output sense.

Next, it is easy to see that VES II as shown in Fig. 5.3 together with the properties of $\Delta u(k)$ and $e'_j(k)$ is totally the same as in the VES II in reference [1].

Then following the similar lines of the proof of Theorem 4.2 in [1], we obtain the conclusion of Theorem 5.4.

That completes the proof of Theorem 5.4.

Similar to the proof of Theorem 5.4 (thus details are omitted here), we have the following corollaries.

Corollary 5.1 *The proposed WMMAC system with weighting Algorithm 2 is stable and convergent if the following conditions are satisfied:*

(1) *There is a model, say $M_j \in \mathbb{M}$, which is closest to the true plant in the following sense with probability one*

$$\sum_{r=1}^{k} e_j(r)^2 < \sum_{r=1}^{k} e_i(r)^2, \forall k \geq d, i \neq j$$

(2) *Each local controller is well defined such that C_i is stabilizing its corresponding local model M_i, $i = 1, \ldots, j, \ldots N$ and the resulting closed-loop system $\{C_i, M_i\}$ is tracking the reference signal $y_r(k)$;*

(3) *For the model closest to the true plant, i.e.,* M_j, *its output approximates the true plant output in the sense that*

$$\frac{1}{n}\sum_{k=1}^{n}\|e'_j(k)\|^2 = \frac{1}{n}\sum_{k=1}^{n}\|e_j(k) - \omega(k)\|^2 = o\left(\frac{1}{n}\sum_{k=1}^{n}\|\phi(k-d)\|^2\right)$$

Corollary 5.2 *The proposed WMMAC system with weighting Algorithm 2 is stable if the following conditions are satisfied:*

(1) *There is a model, say* $M_j \in \mathbb{M}$, *which is closest to the true plant in the following sense with probability one*

$$\sum_{r=1}^{k} e_j(r)^2 < \sum_{r=1}^{k} e_i(r)^2, \forall k \geq d, i \neq j$$

(2) *Each local controller is well defined such that* C_i *is stabilizing its corresponding local model* M_i, $i = 1, \ldots, j, \ldots N$;

(3) *For the model closest to the true plant, i.e.,* M_j, *its output approximates the true plant output in the sense that*

$$\|e'_j(k)\| = \|e_j(k) - \omega(k)\| \leq M < \infty$$

Corollary 5.3 *The proposed WMMAC system with weighting Algorithm 2 is stable if the following conditions are satisfied:*

(1) *There is a model, say* $M_j \in \mathbb{M}$, *which is closest to the true plant in the following sense with probability one*

$$\sum_{r=1}^{k} e_j(r)^2 < \sum_{r=1}^{k} e_i(r)^2, \forall k \geq d, i \neq j$$

(2) *Each local controller is well defined such that* C_i *is stabilizing its corresponding local model* M_i, $i = 1, \ldots, j, \ldots N$;

(3) *For the model closest to the true plant, i.e.,* M_j, *its output approximates the true plant output in the sense that*

$$\frac{1}{n}\sum_{k=1}^{n}\|e'_j(k)\|^2 = \frac{1}{n}\sum_{k=1}^{n}\|e_j(k) - \omega(k)\|^2 \leq M < \infty$$

5.5.3 Extended Results for Parameter Jumping System

Consider the following parameter jumping system

$$P(k): A(q^{-1}, k)y(k) = q^{-d}B(q^{-1}, k)u(k) + \omega(k) \tag{5.39}$$

where

$$A(q^{-1}, k) = 1 + a_1(k)q^{-1} + \cdots + a_{na}(k)q^{-na} \tag{5.40}$$

$$B(q^{-1}, k) = b_0(k) + b_1(k)q^{-1} + \cdots + b_{nb}(k)q^{-nb} \tag{5.41}$$

$$d \geq 1; na \geq 1; nb \geq 1$$

Suppose the system satisfy the following assumptions.

(A1) The parameters $a_i(k), i \in 0, \ldots, n_a$ and $b_i(k), i \in 0, \ldots, n_b$ are piecewise constant functions of k, and the intervals between consecutive parameters jumps are long enough;

(A2) The parameters vector $[a_1(k), \ldots, a_{na}(k); b_0(k), \ldots, b_{nb}(k)]^T$ lies in a limited set.

To deal with the WMMAC of such kind of systems, there are two methods to avoid the zero lock-in condition of p_i from occurring. One method is to set lower bounds for weights $p_i, i = 1, 2, \ldots, N$ as in reference [20]; The other is to reset each weight according to Eq. (5.16) when the parameters jump is detected. In this chapter, we adopted the latter method.

We have the following result regarding the stability of the WMMAC system of parameter jumping plant $P(k)$.

Theorem 5.5 *The proposed WMMAC system with weighting Algorithm 2 is stable and convergent if the following conditions are satisfied:*

(1) Assumptions (A1) and (A2) are satisfied;

(2) At each parameters jump, the true model of plant, M_j, is included in the model set $\mathbb{M}$, which generates the minimum output error in the following sense with probability one

$$\sum_{r=k_\tau}^{k} e_j(r)^2 < \sum_{r=k_\tau}^{k} e_i(r)^2, \forall k \geq k_\tau + d, i \neq j$$

where $k_\tau, \tau = 1, 2, \ldots,$ is the time sequence at which parameters jump occurs.

(3) Each local controller is well defined such that C_i is stabilizing M_i, $i = 1, \ldots, j, \ldots N$, and the output of the resulting closed-loop system $\{C_i, M_i\}$, say $y_{out}(k)$, is tracking the reference signal $y_r(k)$ in the sense that

$$\lim_{n\to\infty} \frac{1}{n} \sum_{k=1}^{n} [y_{out}(k) - y_r(k)]^2 = R', R \le R' < \infty$$

Proof Suppose the jumping parameters are 'frozen', then according to Theorem 5.3, the corresponding WMMAC system is stable and convergent. Further, under Assumptions (A1) and (A2) we know that the over all WMMAC system is a 'slow' switched system between a limited number of above-mentioned stable component systems. Then according to the theory of switched systems [21–25], we obtain the conclusion of Theorem 5.5.

That completes the proof of Theorem 5.5.

To be specific, we need to give a lower bound of admissible intervals between consecutive parameters jumps for the WMMAC scheme to remain stable. To this end, we construct the VES III for the parameter jump situation based on Condition (2), as shown in Fig. 5.4. Note that in VES III, j is not a fixed value.

We have the following result regarding parameter jumping plant.

Theorem 5.6 *The proposed WMMAC system with weighting Algorithm 2 is stable if the following conditions are satisfied:*

(1) Assumption (A2) is satisfied, and the intervals between consecutive parameters jumps are not smaller than "dwell-time" $\tau_D^{'}$

$$\tau_D^{'} = \max\{d - d \ln N + L \ln 10, \tau_D\} \tag{5.42}$$

where d is the time delay as shown in (5.2) and (5.39), N is the number of local controllers, 10^{-L} *is the smallest value representable in a specific floating-point system,* τ_D *is the "dwell-time" of VES III (as shown in Fig. 5.4) with* $\Delta u(k) = 0$*, see Appendix B for its calculation;*

(2) At each parameters jump, the true model of plant, M_j*, is included in the model set* $\mathbb{M}$*, which generates the minimum output error in the following sense with probability one*

$$\sum_{r=k_\tau}^{k} e_j(r)^2 < \sum_{r=k_\tau}^{k} e_i(r)^2, \forall k \ge k_\tau + d, i \ne j$$

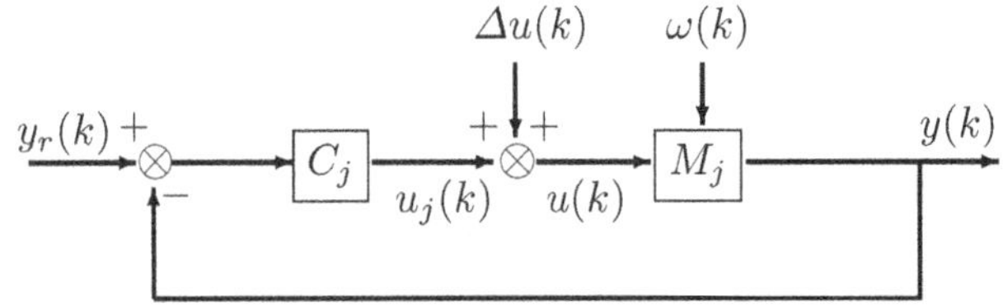

Fig. 5.4 Virtual equivalent system III of WMMAC system: parameter jumping

where $k_\tau, \tau = 1, 2, \ldots,$ is the time sequence at which parameters jump occurs.

(3) *Each local controller is well defined such that C_i is stabilizing M_i, $i = 1, \ldots, j, \ldots N$, and the output of the resulting closed-loop system $\{C_i, M_i\}$, say $y_{out}(k)$, is tracking the reference signal $y_r(k)$ in the sense that*

$$\lim_{n\to\infty} \frac{1}{n} \sum_{k=1}^{n} [y_{out}(k) - y_r(k)]^2 = R', R \le R' < \infty$$

Proof Step 1
We first consider the convergence rate of weighting Algorithm 2.

According to Condition (2) and Lemma B.2 in Appendix B we know that $[\beta_i(k)]^{ceil\left(\frac{1}{1-\beta_i(k)}\right)}, i \neq j$ is monotonically increasing as $\beta_i(k)$ approaches 1 from the left, and that

$$[\beta_i(k)]^{ceil\left(\frac{1}{1-\beta_i(k)}\right)} \le \frac{1}{e}, k \ge d, i \neq j \tag{5.43}$$

then we have

$$l_i(k) \le \left(\frac{1}{N}\right)^d \left(\frac{1}{e}\right)^{k-d}, i \neq j \tag{5.44}$$

Considering the limited precision of a specific floating-point system, i.e., $10^{-L} \backsimeq 0$, then we claim that $l_i(k)$ will be zero within K steps

$$\left(\frac{1}{N}\right)^d \left(\frac{1}{e}\right)^{K-d} = 10^{-L} \tag{5.45}$$

$$K = d - d \ln N + L \ln 10 \tag{5.46}$$

Then we have $u(k) = 0, \forall k_\tau + K \le k \le k_{\tau+1}$, where $k_\tau, \tau = 1, 2, \ldots,$ is the parameter jump sequence.

Thus $\Delta u(k)$ in VES III (Fig. 5.4) is a bounded signal.

Step 2
VES III (Fig. 5.4) can be decomposed into two subsystems, as shown in Figs. 5.5 and 5.6, respectively. Note that $y(k) = 0$, $u(k) = 0$, $\omega(k) = 0$, $\Delta u(k) = 0$, $y_1(k) = 0$, $u_1(k) = 0$, $y_2(k) = 0$, $u_2(k) = 0$ for $k \le 0$. Then we have

$$u(k) = u_1(k) + u_2(k) \tag{5.47}$$

$$y(k) = y_1(k) + y_2(k) \tag{5.48}$$

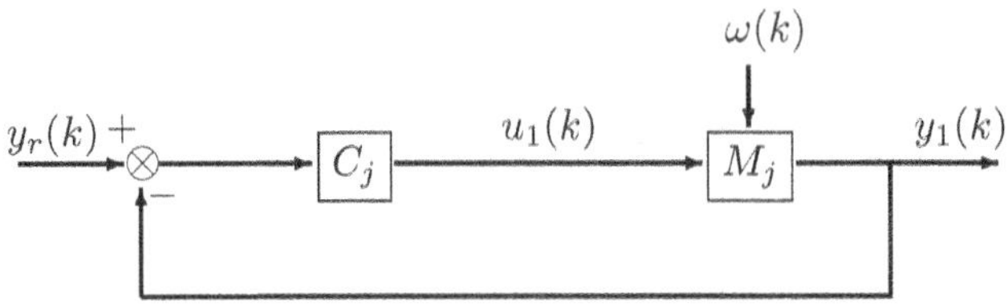

Fig. 5.5 Subsystem 1 of VES III

Step 3
Subsystem 1, as shown in Fig. 5.5, is a switching system between N component LTI systems. According to Condition (3) we know that each component LTI system is stable. Then with appropriate "dwell time" τ_D, subsystem 1 of VES III, as shown in Fig. 5.5, is stable and tracking, see Appendix B for details of the calculation of τ_D. Then we have

$$\lim_{n\to\infty}\frac{1}{n}\sum_{k=1}^{n}[y_1(k)]^2 \le M_1 < \infty,\ \lim_{n\to\infty}\frac{1}{n}\sum_{k=1}^{n}[u_1(k)]^2 \le M_2 < \infty$$

$$\lim_{n\to\infty}\frac{1}{n}\sum_{k=1}^{n}[y_1(k)-y_r(k)]^2 = \lim_{n\to\infty}\frac{1}{n}\sum_{k=1}^{n}[y_{out}(k)-y_r(k)]^2 = R',\ R \le R' < \infty$$

Step 4
Subsystem 2, as shown in Fig. 5.6, is identical to subsystem 1 in structure. Since $\omega(k) = 0$ in subsystem 2, it is a stable deterministic system with reference signal $y_r(k) = 0$, and bounded "disturbance" $\Delta u(k)$. Then we have

$$\lim_{n\to\infty}\frac{1}{n}\sum_{k=1}^{n}[y_2(k)]^2 \le M_3 < \infty,\ \lim_{n\to\infty}\frac{1}{n}\sum_{k=1}^{n}[u_2(k)]^2 \le M_4 < \infty$$

Thus, combining with (5.47) and (5.49) we know that input-output signals, i.e, $u(k)$ and $y(k)$ in VES III are bounded.

Further we have

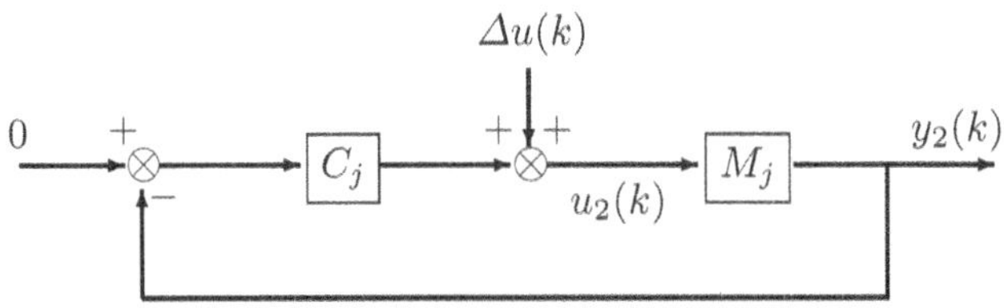

Fig. 5.6 Subsystem 2 of VES III

$$\lim_{n\to\infty}\frac{1}{n}\sum_{k=1}^{n}[y(k)-y_r(k)]^2 = R'' < \infty$$

where R'' is estimated as follows.

Consider

$$\frac{1}{n}\sum_{k=1}^{n}[y_1(k)-y_r(k)+y_2(k)]^2 = \frac{1}{n}\sum_{k=1}^{n}[y_1(k)-y_r(k)]^2 + \frac{1}{n}\sum_{k=1}^{n}[y_2(k)]^2 + \frac{2}{n}\sum_{k=1}^{n}[y_1(k)-y_r(k)]y_2(k) \tag{5.49}$$

By the fact $2ab \le a^2+b^2$ we have

$$\frac{2}{n}\sum_{k=1}^{n}[y_1(k)-y_r(k)]y_2(k) \le \frac{1}{n}\sum_{k=1}^{n}[y_1(k)-y_r(k)]^2 + \frac{1}{n}\sum_{k=1}^{n}[y_2(k)]^2 \tag{5.50}$$

Thus

$$\begin{aligned}
&\lim\frac{1}{n}\sum_{k=1}^{n}[y_1(k)-y_r(k)+y_2(k)]^2\\
&\le 2\left\{\lim\frac{1}{n}\sum_{k=1}^{n}[y_1(k)-y_r(k)]^2 + \lim\frac{1}{n}\sum_{k=1}^{n}[y_2(k)]^2\right\}\\
&= 2(R'+M_3)
\end{aligned} \tag{5.51}$$

That completes the proof of Theorem 5.6.

5.6 Simulation Results

To verify the effectiveness and the correctness of the proposed weighting algorithm, the stability and dynamic performance of the corresponding closed-loop WMMAC system, this section presents some simulation results conducted with MATLAB® version 6.5.

Consider an uncertain discrete-time plant

$$(1+a_1q^{-1}+a_2q^{-2})y(k) = q^{-1}(b_0+b_1q^{-1})u(k)+\omega(k) \tag{5.52}$$

where $\omega(k)$ is a zero-mean white noise sequence with variance $\sigma = 0.1$ that was created with the **randn()** function. The deterministic part of Eq. (5.53) is obtained by converting the following continuous-time LTI model to a discrete-time model with sample time $t_s = 0.5s$ and the zero order hold.

$$\frac{k}{s^2 - 3s + 2} \tag{5.53}$$

The uncertainties of Eq. (5.53) originate from k in Eq. (5.54). For the real plant, we suppose $k = 1$. For simplicity, the model set includes four models, corresponding to $k = k_1$, $k = k_2$, $k = k_3$, and $k = k_4$, respectively. Four 'local' controllers were designed by pole assignment strategy. Each controller stabilizes a possible model by formulating an expected closed-loop characteristic polynomial, say $A_m(q^{-1})$ and track the reference signal $y_r(k)$.

$$A_m(q^{-1}) = 1 - 1.3205q^{-1} + 0.4966q^{-2} \tag{5.54}$$

which corresponds to the characteristic polynomial of the following continuous time second-order system

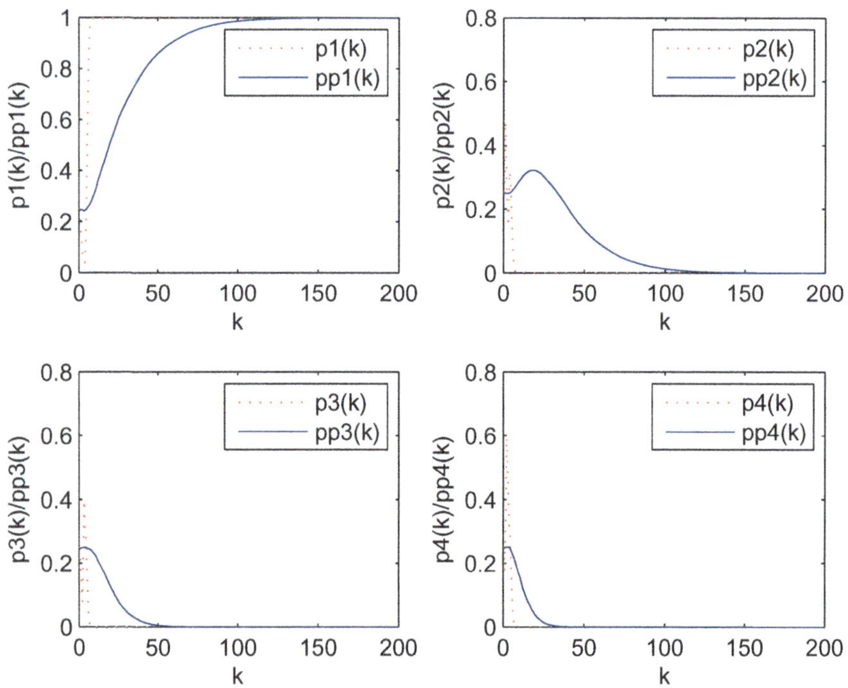

Fig. 5.7 Weight signals of case 1, pi: produced by weighting Algorithm 2; ppi: produced by weighting Algorithm 1

$$\frac{\omega_n^{\ 2}}{s^2 + 2\xi\omega_n s + \omega_n^{\ 2}} \tag{5.55}$$

with $\xi = 0.707$, $\omega_n = 1$, and sampling time $t_s = 0.5\,\mathrm{s}$.

Case 1: $k_1 = 1.01, k_2 = 1.02, k_3 = 1.03, k_4 = 1.04$. The simulation results, i.e., the four weights signals, the closed-loop output $y(k)$ against reference signal $y_r(k)$, and the control signal $u(k)$ are shown in Figs. 5.7 and 5.8, respectively.

Case 2: $k_1 = 1.01, k_2 = 1.010000000000001, k_3 = 1.03, k_4 = 1.04$. The simulation results are shown in Figs. 5.9 and 5.10, respectively.

In Case 1, with both weighting algorithms, four controller weights $p_i(k), i = 1, 2, 3, 4$ converge correctly and consequently the closed-loop control system is stable and convergent.

In Case 2, weighting Algorithm 1 can not yields correct convergence of weights $p_i(k), i = 1, 2, 3, 4$, while weighting Algorithm 2 can still converge correctly.

In both cases, weighting Algorithm 2 converges faster than weighting Algorithm 1. This coincides with the theoretical analysis.

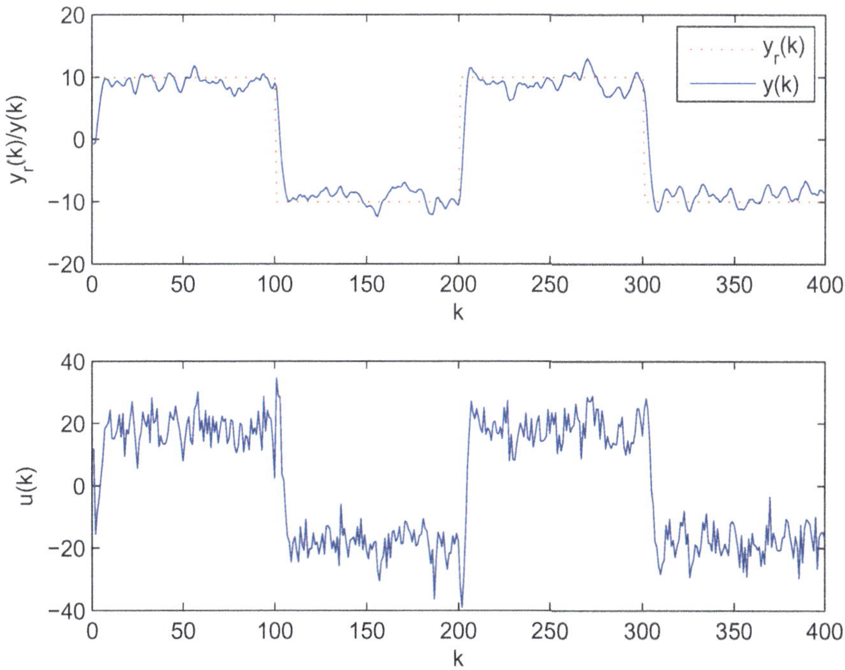

Fig. 5.8 Output, reference, and control signals of case 1

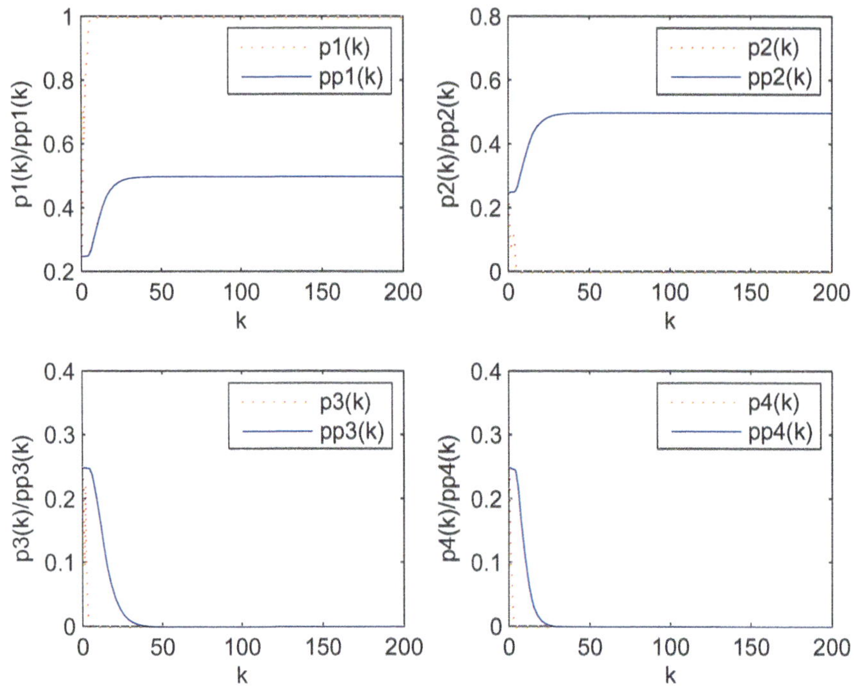

Fig. 5.9 Weight signals of case 2, pi: produced by weighting Algorithm 2; ppi: produced by weighting Algorithm 1

5.7 Conclusion

An improved weighting algorithm was proposed with convergence analysis result under the assumption that local model output error indexes are discriminable. Based on the convergence of the weighting algorithm, the stability and convergence of corresponding WMMAC system have been proved with the help of VES concept and methodology.

To some extent, the stability and convergence analysis methodology based on VES is independent of specific 'local' control strategy and specific weighting algorithm. Actually, the stability and convergence of the closed-loop WMMAC system depend only on three conditions, i.e., the stabilizing characteristic of each 'local' controller; the effective model set; and the appropriate convergence of the weighting algorithm. Thus, WMMAC provide us with more confidence and flexibility in adaptive control engineering.

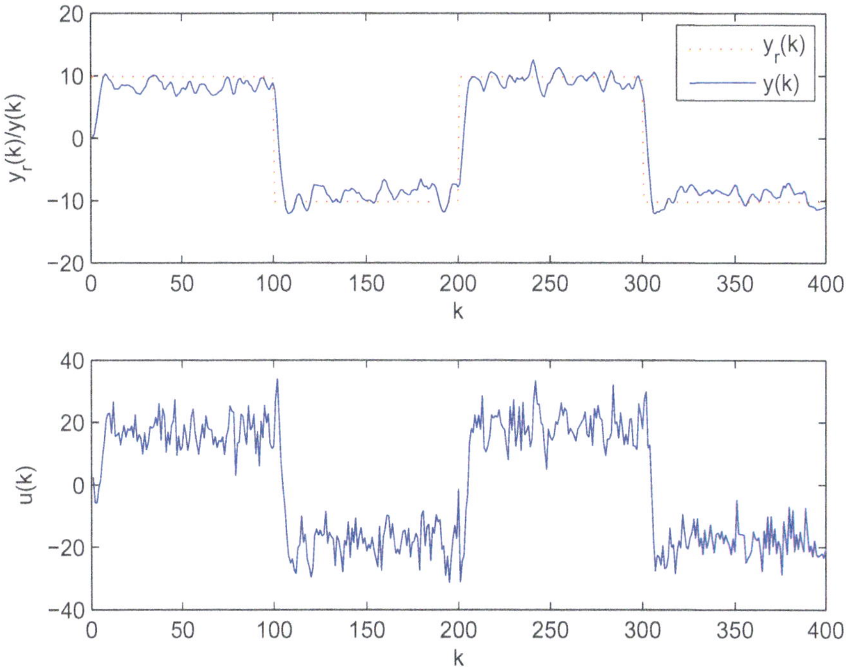

Fig. 5.10 Output, reference, and control signals of case 2

References

1. W. Zhang, Stable weighted multiple model adaptive control: discrete-time stochastic plant. Int. J. Adapt. Control Signal Process **27**(7), 562–581 (2013)
2. D.T. Magill, Optimal adaptive estimation of sampled stochastic processes. IEEE Trans. Autom. Control **10**, 434–439 (1965)
3. D.G. Lainiotis, Partitioning: aunifying framework for adaptive systems I: Estimation; II: Control, in *Proceedings of IEEE*, vol. 64 (1976), pp. 1126–1143 and 1182–1197
4. M. Athans et al., The stochastic control of the F-8C aircraft using a multiple model adaptive control (MMAC) method-Part I: Equilibrium flight. IEEE Trans. Autom. Control **22**, 768–780 (1977)
5. D.W. Lane, P.S. Maybeck, Multiple model adaptive estimation applied to the Lambda URV for failure detection and identification, in *Proceedings IEEE 33rd Conference Decision Control* (Lake Buena Vista, FL, 1994), pp. 678–683
6. C. Yu, R.J. Roy, H. Kaufman, B.W. Bequette, Multiple-model adaptive predictive control of mean arterial pressure and cardiac output. IEEE Trans. Biomed. Eng. **39**, 765–778 (1992)
7. R.L. Moose, H.F. Van Landingham, D.H. McCabe, Modeling and estimation for tracking maneuvering targets, in IEEE Trans. Aerospace Elec. Syst. **AES-15**, 448–456 (1979)
8. X.R. Li, Y. Bar-Shalom, Design of an interacting multiple model algorithm for air traffic control tracking. IEEE Trans. Control Syst. Tech. **1**, 186–194 (1993)
9. M. Athans, S. Fekri, A. Pascoal, Issues on robust adaptive feedback control, in *Preprints 16th IFAC World Congress, Invited Plenary paper* (Prague, Czech Republic, 2005), pp. 9–39
10. S. Fekri, M. Athans, A. Pascoal, Issues, progress and new results in robust adaptive control. Int. J. Adapt. Control Signal Process **20**(10), 519–579 (2006)

11. S. Fekri, M. Athans, A. Pascoal, Robust multiple model adaptive control (RMMAC): a case study. Int. J. Adapt. Control Signal Process **21**(1), 1–30 (2007)
12. Y. Baram, Information, consistent estimation and dynamic system identification, Ph.D. Dissertation (MIT, Cambridge, MA, USA., 1976)
13. Y. Baram, N.R. Sandell, An information theoretic approach to dynamical systems modeling and identification. IEEE Trans. Autom. Control **23**(1), 61–66 (1978)
14. Y. Baram, N.R. Sandell, Consistent estimation on finite parameter sets with application to linear systems identification. IEEE Trans. Autom. Control **23**(3), 451–454 (1978)
15. A. Kehagias, Convergence properties of the lainiotis partition algorithm. Control Comput. **19**, 1–6 (1991)
16. M. Kuipers, P. Ioannou, Practical robust adaptive control: benchmark example, in *Proceedings of American Control Conference* (Seattle, Washington, USA, 2008)
17. M. Kuipers, P. Ioannou, Multiple model adaptive control with mixing. IEEE Trans. Autom. Control **55**(8), 1822–1836 (2010)
18. N. Sadati, G.A. Dumont, H.R. Feyz Mahdavian, Robust multiple model adaptive control using fuzzy fusion, in *42nd South Eastern Symposium on System Theory* (Tyler, TX, USA, 2010)
19. W. Zhang, On the stability and convergence of self-tuning control-virtual equivalent system approach. Int. J. Control **83**(5), 879–896 (2010)
20. J.A. Custafson, P.S. Maybeck, Flexible spacestructure control via moving-bank multiple model algorithms. IEEE Trans. Aerospace Electr. Syst. **30**(3), 750–757 (1994)
21. D. Liberzon, A.S. Morse, Basic problems in stability and design of switched systems. IEEE Control Syst. Mag. **19**, 59–70 (1999)
22. D. Chatterjee, D. Liberzon, Stability analysis of deterministic and stochastic switched systems via a comparison principle and multiple Lyapunov functions. SIAM J. Control Optim. **45**(1), 174–206 (2006)
23. J.P. Hespanha, A.S. Morse, Switching between stabilizing controllers. Automatica **38**, 1905C1917 (2002)
24. H. Lin, P.J. Antsaklis, Stability and stabilizability of switched linear systems: a survey of recent results. IEEE Trans. Aerospace Electr. Syst. **54**(2), 308–322 (2009)
25. R. Shorten, F. Wirth, O. Mason, K. Wulff, C. King, Stability criteria for switched and hybrid systems. SIAM Rev. **49**(4), 545–592 (2007)

Chapter 6
Stable Weighted Multiple Model Adaptive Control of Continuous-Time Plant

Abstract This chapter deals with the stability of WMMAC of continuous-time plant with large parametric uncertainties, which is not a trivial problem although the stability of discrete-time WMMAC systems has been proved in Chaps. 4 and 5 for different situations. In continuous-time WMMAC system, each 'local' controller is designed by the mixed-μ-synthesis method to deal with small parametric uncertainties of plant together with disturbance, the weighting algorithm is formulated directly based on local model output errors rather than the residuals generated by multiple Kalman filters as in classical multiple model adaptive control (CMMAC). The closed-loop stability (signal boundedness) and tracking performance of the proposed WMMAC system are proved with the help of VES concept and methodology.

6.1 Introduction

A lot of achievements have been made on the road towards "robust" and "adaptive" characteristics in control theory and applications. Among others, the robust multiple model adaptive control (RMMAC) architecture provides an attractive framework for searching the "holy grail" of robust adaptive control [1], which inherits the basic structure of CMMAC [2–5] with some improvements. Roughly speaking, RMMAC differs from CMMAC mainly in three aspects:

(1) As local control strategy, output dynamic compensator by mixed-μ-synthesis method replaced LQ state feedback control strategy in CMMAC scheme;
(2) In RMMAC scheme, local control and Kalman filter are separated, Kalman filters are only used for state estimation (also known as multiple model adaptive estimation) and model identification (also known as weighting algorithm or posterior probability evaluator: PPE);
(3) Performance-driven methodology to determine the number of required models.

W. Zhang and Q. Li, *Virtual Equivalent System Approach for Stability Analysis of Model-based Control Systems*, https://doi.org/10.1007/978-981-15-5538-1_6

In Reference [6], Kuipers and Ioannou pointed out that the RMMAC scheme outperforms the best achievable non-adaptive scheme in the well-known bench-mark problem, i.e., the two-cart mass-spring-damper (MSD) system [1]. However RMMAC scheme might have poor performance due to either large initial state estimates error or inaccurate knowledge of the disturbance/noise statistics. Additionally, the complexity and calculation burden of RMMAC may hinder its engineering application because every candidate local controller requires a corresponding Kalman filter and a weighting algorithm or post posterior evaluation. Consequently, some other modified RMMAC schemes have been proposed without Kalman filters, such as adaptive mixing control (AMC) [6–8] and fuzzy-fusion [9] based RMMAC schemes. These modified schemes have demonstrated some merits from different aspects. Fuzzy-fusion based scheme is very simple, but the membership function needs to be determined off-line, which may bring in some inaccuracies. AMC approach was proposed with stability and robustness analysis, but the requirements on the plant information, the estimator convergence, and the mixer performance, are still too strict to be applicable to practical situations. Among others, the well-known persistent excitation (PE) condition is still required in the estimator of AMC, which could be inconvenience in real engineering applications.

A novel WMMAC scheme was presented in reference [10], which makes weighted sum directly on the multiple model parameter vectors as plant parameters estimates. Compared to classical adaptive control schemes, this approach has faster estimates convergence rate, resulting in better closed-loop performance. However, this approach requires that the parameter vector of the plant lies in a convex hull of the parameter vectors of multiple models. Besides, the standard assumptions for model reference adaptive control are also required in this approach [11].

In references [12, 13], a new weighting strategy (algorithm) for WMMAC of discrete-time stochastic plant, is proposed, in which each 'local' controller can be designed by any existing control strategies, such as pole placement, etc, if only each 'local' controller stabilizes its corresponding 'local' model. In addition, the closed-loop stability of the resulting WMMAC system is proved based on VES concept and methodology [14].

According to references [6, 12, 13], an improved WMMAC scheme is proposed in this Chapter for continuous-time plant with large parametric uncertainties, modeling error, and bounded disturbance. In detail, the proposed WMMAC scheme consists of dynamic output feedback compensator designed by mixed-μ-synthesis method or any other robust control strategies, and the weighting algorithm proposed in [13] with modification to be suitable for continuous-time system.

6.2 Description of the WMMAC System

Consider the following single-input-single-output (SISO) plant P with large parametric uncertainties.

$$\begin{cases} \dot{x}_p(t) = A(\theta)x_p(t) + B(\theta)u(t) \\ y_p(t) = D(\theta)x_p(t) + E(\theta)u(t) + d(t) \\ x_p(0) = 0 \end{cases} \tag{6.1}$$

where $u(t)$, $y_p(t)$, and $x_p(t)$ are input, output, and state vectors of the plant, respectively; $A(\theta)$, $B(\theta)$, $D(\theta)$, and $E(\theta)$ are vectors or matrices with proper dimensions; θ denotes the parametric uncertainties, which are supposed to belong to a limited set $\Theta = \{\theta_1, \theta_2, \ldots, \theta_m\}$; $d(t)$ represents bounded disturbance.

To design a WMMAC system, there are mainly three tasks, i.e., (1) to select a model set (it is a modeling problem); (2) to design the 'local' controllers set; (3) to construct a weighting algorithm. First of all, a model set should be established to cover the uncertainties of plant, i.e., $\mathbb{M} = \{A(\theta), B(\theta), D(\theta), E(\theta)\}$, or for simplicity, denote $\mathbb{M} = \{M_i\} = \{A_i, B_i, D_i, E_i\}$, $i = 1, 2, \ldots, m$. Each 'local' model M_i has the following state-space realization

$$\begin{cases} \dot{x}_{mi}(t) = A_i x_{mi}(t) + B_i u(t) \\ y_{mi}(t) = D_i x_{mi}(t) + E_i u(t) + d(t) \\ x_{mi}(0) = 0 \end{cases} \tag{6.2}$$

Then for each local model M_i with smaller parametric uncertainties, a corresponding local robust controller or local robust compensator C_i is designed by mixed-μ-synthesis method (details are omitted here, since it can be readily done based on Matlab functions). The objective of each local controller is to track a known and bounded reference signal, i.e., $y_r(t)$, $|y_r(t)| \le \eta < \infty$.

To be specific, each local robust controller C_i is an admissible controller that, together with M_i, constitutes an internal stable closed-loop system as shown in Fig. 6.1, in which $C_i(s)$ and $M_i(s)$ are transfer functions of C_i and M_i, respectively.

Local control signal $u_i(t)$ is generated by local controller C_i with the following state-space realization

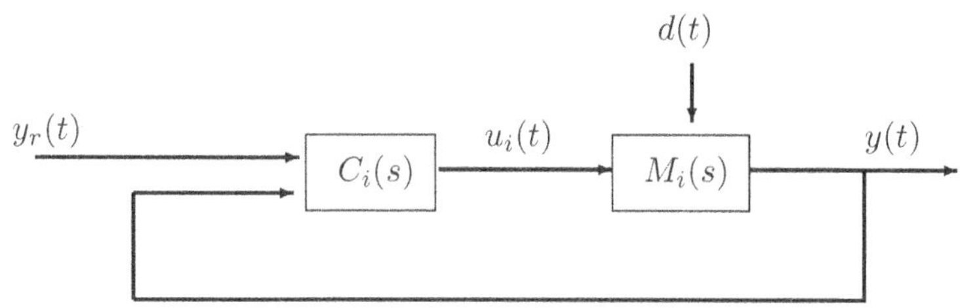

Fig. 6.1 Design diagram of a local robust controller

$$\begin{cases} \dot{x}_{ci}(t) = A_{ci}x_{ci}(t) + B_{ci}y(t) \\ u_i(t) = K_i x_{ci}(t) + G_i y_r(t) \\ x_{ci}(0) = 0 \end{cases} \tag{6.3}$$

where K_i represents the feedback gain, G_i represents the feed-forward gain.

Accordingly, global control law is obtained by

$$u(t) = \sum_{i=1}^{m} p_i(t)u_i(t) \tag{6.4}$$

where $p_i(t)$ is the weight for 'local controller' C_i.

Based on each local model output error, i.e.

$$e_i(t) \triangleq y_p(t) - y_{mi}(t), i = 1, 2, \ldots, m$$

we adopt the following weighting algorithm, which was originated from reference [13].

$$l_i(0) = \frac{1}{m};\ p_i(0) = l_i(0) \tag{6.5}$$

$$l_i'(k) = \alpha + \frac{1}{t_k}\int_0^{t_k} e_i(t)^2 dt \tag{6.6}$$

$$l_{min}(k) = \min_i l_i'(k) \tag{6.7}$$

$$\beta_i(k) = \frac{l_{min}(k)}{l_i'(k)} \tag{6.8}$$

$$l_i(k) = \begin{cases} l_i(k-1) & \text{if } \beta_i(k) = 1 \\ l_i(k-1)[\beta_i(k)]^{ceil\left(\frac{1}{1-\beta_i(k)}\right)} & \text{if } \beta_i(k) < 1 \end{cases} \tag{6.9}$$

$$p_i(k) = \frac{l_i(k)}{\sum_{i=1}^{m} l_i(k)} \tag{6.10}$$

$$p_i(t) = p_i(k), t_k \le t < t_{k+1} \tag{6.11}$$

where k is the sampling time sequence, $\alpha > 0$, a small constant to avoid $l_i'(k) = 0$; $ceil(x)$, the ceiling function that generates the smallest integer not less than x, i.e.,

$$ceil(x) = \min\{n \in \mathbb{Z} | x \le n\}$$

The earlier version of weighting algorithm (6.5)–(6.10) can be found in reference [12], which is more easier to understand, but its convergence speed is a bit slow. For comparisons, here we give the original weighting algorithm in reference [12].

$$l_i(0) = \frac{1}{m};\ p_i(0) = l_i(0)$$

$$l_i'(k) = 1 + \frac{1}{k}\sum_{r=1}^{k} e_i^2(r)$$

$$l_{min}(k) = \min_i l_i'(k)$$

$$l_i(k) = l_i(k-1)\frac{l_{min}(k)}{l_i'(k)}$$

$$p_i(k) = \frac{l_i(k)}{\sum\limits_{i=1}^{m} l_i(k)}$$

Remark 6.1 The above-mentioned weighting algorithm was motivated by a simple thinking: the closer the model to the plant to be controlled, the smaller the signal energy of the model output error $e_i(k)$. It can also be regarded as an on-line learning algorithm.

As the convergence analysis of weighting algorithm Eqs. (6.5)–(6.11), we give the following analysis result.

Theorem 6.1 *Suppose there is a model, say $M_l \in \mathbb{M}$, which is closest to the true plant in the following sense*

$$\int_0^{t_k} [e_l(t)]^2 dt < \int_0^{t_k} [e_i(t)]^2 dt,\ \forall k \geq k^*, i \neq l \tag{6.12}$$

where k^* is an unknown limited sampling time instant.

Then we get the following convergence result

$$\lim_{t\to\infty} p_l(t) = 1;\ \lim_{t\to\infty} p_i(t) = 0, i \neq l$$

Proof Considering equations (6.5)–(6.9) under condition (6.12), we have the following facts:

$$\begin{cases} l_{min}(k) = l_l'(k) \\ \frac{l_{min}(k)}{l_l'(k)} = 1 & \forall k \geq k^*, i \neq l \\ \frac{l_{min}(k)}{l_i'(k)} < 1 \end{cases}$$

Further we obtain

$$\beta_l(k) = 1, \beta_i(k) < 1, i \neq l, \forall k \geq k^*$$

$$l_l(k) = l_l(k^*), l_i(k) \to 0, i \neq l, \forall k \geq k^*$$

For the only exception situation $\beta_i(k) < 1$, $\beta_i(k) \to 1$ we still have

$$\lim_{k\to\infty} [\beta_i(k)]^{ceil(\frac{1}{1-\beta_i(k)})} = \frac{1}{e} < 1 \tag{6.13}$$

it is because, in Eq. (6.9, the sequence $[\beta_i(k)]^{ceil(\frac{1}{1-\beta_i(k)})}$ is monotonically increasing with upper bound $1/e$ (see Lemma A1 in Reference [13] for details).

Then by Eq.(6.10), we have

$$p_i(k) \to 0, p_l(k) \to 1, i \neq l$$

Further consider the zero-order-hold (ZOH), i.e., Eq. (6.11), we have

$$\lim_{t\to\infty} p_l(t) = 1; \lim_{t\to\infty} p_i(t) = 0, i \neq l$$

That completes the proof. □

Up to now, we can construct a concise block diagram of the WMMAC system, as shown in Fig. 6.2, in which the details of weighting algorithm are omitted to save space.

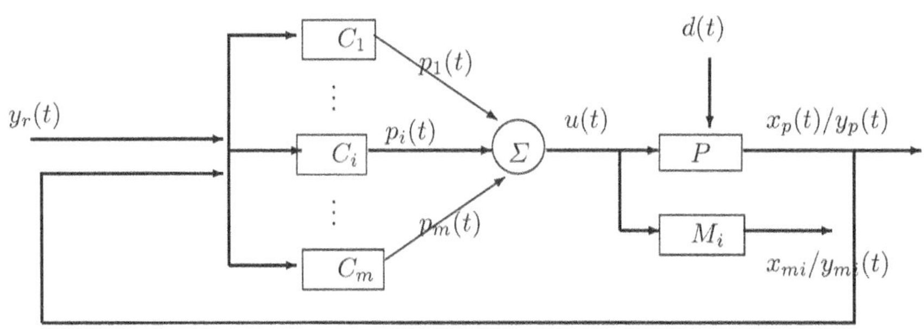

Fig. 6.2 Block diagram for the WMMAC

Next, the main results of the proposed WMMAC scheme will be developed based on Fig. 6.2 and Theorem 6.1.

6.3 Main Results

In this section, the stability and tracking performance analysis of the proposed WMMAC system will be given with the help of VES concept and methodology. We will consider two situations, situation 1: the true model of the plant is included in the model set; situation 2: the true model is not included in the model set, but there is a closest model to the plant with bounded or vanishing modeling errors. For situation 1, we have Theorem 6.2; for situation 2, we have Corollaries 6.1 and 6.2.

Theorem 6.2 *The proposed WMMAC system for plant described in Eq. (6.1) with weighting algorithm (6.5)–(6.11) is stable and tracking the reference signal, if only the following conditions are satisfied:*

(1) The true model of plant, denoted as M_l, is included in the model set $\mathbb{M}$, and condition (6.12) is satisfied;
(2) Each local controller C_i is stable itself, C_i and M_i constitute an internal stable closed-loop system whose output is tracking the reference signal $y_r(t)$;
(3) Disturbance $d(t)$ is bounded and tends to zero.

Proof We will prove Theorem 6.2 in two cases, i.e., Case 1: $d(t) = 0$; and Case 2: $d(t) \to 0$.

First we consider Case 1 in 4 steps.

Step 1: The convergence of weighting algorithm.
According to Theorem 6.1, condition (1) of Theorem 6.2 yields

$$p_l(t) \to 1;\ p_i(t) \to 0, i \neq l$$

Step 2: The construction of VES and the property of $\Delta u(t)$.
The original WMMAC system as shown in Fig. 6.2 is equivalent, in the input-output sense, to the VES as shown in Fig. 6.3.

In Fig. 6.3, $\Delta u(t) = u(t) - u_l(t)$ is a compensation signal. It is worth pointing out that from Figs. 6.2 and 6.3, a nonlinear system in structure is converted into a linear system in structure with a nonlinear compensation signal $\Delta u(t)$. In other words, a nonlinear dominant (in structure) problem is converted into a linear dominant (in structure) problem. The nonlinearity in structure of the original system has been transferred to a 'small' nonlinear signal (i.e. $\Delta u(t)$) in an LTI system that is easier to deal with.

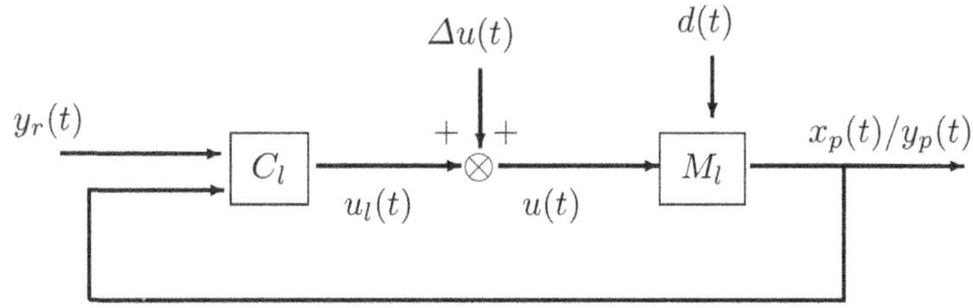

Fig. 6.3 Block diagram of the VES for the WMMAC

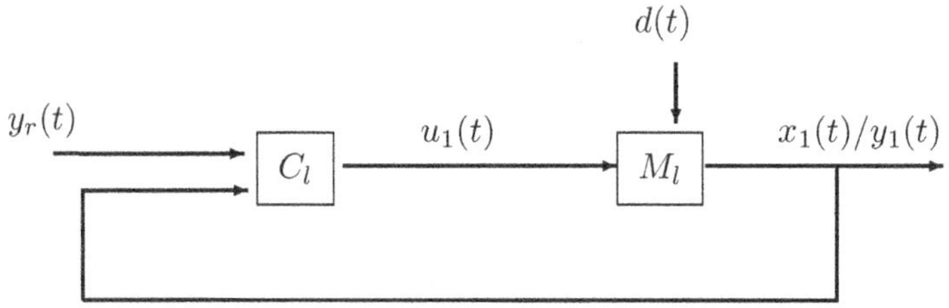

Fig. 6.4 Block diagram for decomposed subsystem I of the VES (Fig. 6.3)

Considering Condition (2), i.e., each local controller is stable, then by Lemma C.1, we have

$$||x_{ci}(t)|| = O(||x_{cl}(t)||), i = 1, 2, \ldots, m, i \neq l \tag{6.14}$$

Further considering $p_l(t) \to 1$, $p_i(t) \to 0, i \neq l$, we get the property of $\Delta u(t)$

$$\begin{aligned} \|\Delta u(t)\| &= \|u(t) - u_l(t)\| \\ &= \|\sum_{i=1}^{m} p_i(t)[K_i x_{ci}(t) + G_i y_r(t)] - K_l x_{cl}(t) - G_l y_r(t)\| \\ &= o\Big(\sum_{i=1}^{m} ||x_{ci}(t)|| + \eta\Big) \\ &= o(\|x_{cl}(t)\| + \eta) \end{aligned} \tag{6.15}$$

where $|y_r(t)| \leq \eta < \infty$.

Step 3: Decomposition of the VES
Since Fig. 6.3 is an LTI system with zero initial state and compensation signal $\Delta u(t)$, it can be decomposed into two subsystems with zero initial states, i.e., subsystem I as shown in Fig. 6.4, and subsystem II as shown in Fig. 6.5, respectively.

Actually, subsystem I, as shown in Fig. 6.4, is the 'ideal' closed-loop control system for the WMMAC system, i.e., the convergence target for the WMMAC system.

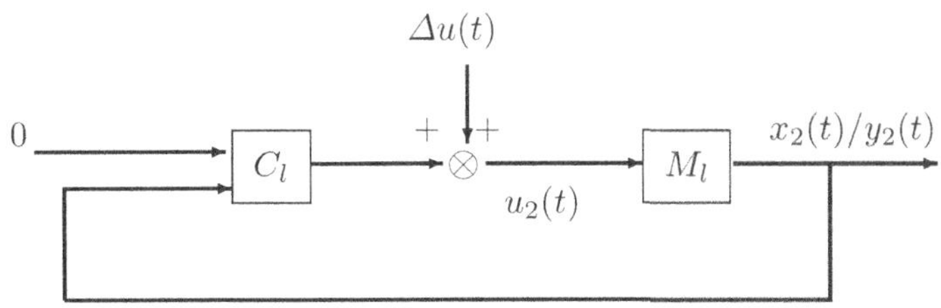

Fig. 6.5 Block diagram for decomposed subsystem II of the VES (Fig. 6.3)

Denote

$$x(t) = \left[x_p(t)\ x_{cl}(t)\right]^T$$

$$x_1(t) = \left[x_{1p}(t)\ x_{1cl}(t)\right]^T$$

$$x_2(t) = \left[x_{2p}(t)\ x_{2cl}(t)\right]^T$$

as the augmented state vectors of the VES (Fig. 6.3), subsystem I (Fig. 6.4), and subsystem II (Fig. 6.5), respectively. For simplicity, the details of state-space descriptions of these systems are omitted here, but we know that they are all internal stable LTI systems.

According to linear superposition principle, we obtain

$$\begin{aligned} x(t) &= x_1(t) + x_2(t) \\ y(t) &= y_1(t) + y_2(t) \end{aligned} \tag{6.16}$$

Step 4: The proof of the conclusion.
By condition (2), subsystem I as shown in Fig. 6.4 is an internal stable LTI system, and its output is tracking the reference signal, i.e.

$$||x_1(t)|| < \infty,\ y_1(t) \to y_r(t) \tag{6.17}$$

Obviously, subsystem II as shown in Fig. 6.5 is also an internal stable LTI system with reference signal $y_r(t) = 0$, $\Delta u(t)$ satisfying (6.15). Further noticing $\|x_{cl}(t)\| \leq \|x(t)\|$ by definition, we get the following result according to Lemma C.2 and Corollary C.1 in Appendix C

$$\begin{aligned} \|x_2(t)\| &= o(\|x_{cl}(t)\| + \eta) \\ &= o(\|x(t)\| + \eta) \end{aligned} \tag{6.18}$$

Putting (6.16), (6.17), and (6.18) together, according to Lemma C.3, we obtain

$$\|x(t)\| < \infty$$

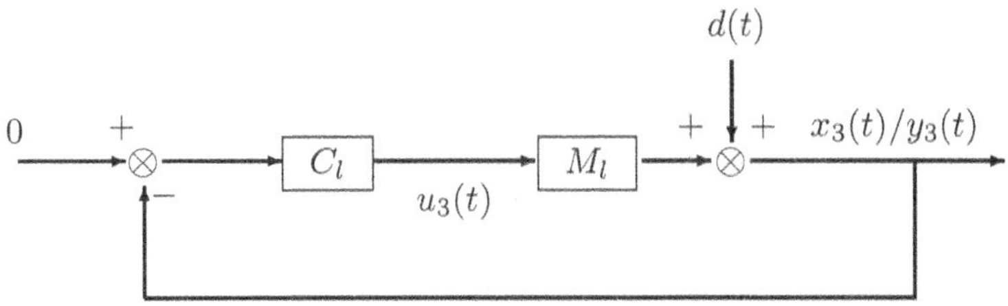

Fig. 6.6 Block diagram for decomposed subsystem III of the VES (Fig. 6.3): $d(t) \to 0$

Thus, by (6.18) we have

$$\|x_2(t)\| \to 0, x_2(t) \to 0, y_2(t) \to 0$$

Consequently we get

$$y(t) = y_1(t) + y_2(t) \to y_1(t) \to y_r(t) \tag{6.19}$$

Case 2: $d(t) \to 0$
Next we consider the non-ideal situation, the disturbance is not zero but bounded and vanishing, i.e., $d(t) \to 0$. Obviously disturbance $d(t)$ can also be treated as $\Delta u(t)$, and accordingly we have the third decomposed subsystem as shown in Fig. 6.6. Then by linear superposition principle we have

$$\begin{aligned} x(t) &= x_1(t) + x_2(t) + x_3(t) \\ y(t) &= y_1(t) + y_2(t) + y_3(t) \end{aligned} \tag{6.20}$$

It is not difficult to see that the properties of $x_1(t)$, $x_2(t)$ remain unchanged as in Case 1. Since subsystem III is a stable system with input signal $d(t) \to 0$, then it is not difficult to see that

$$x_3(t) \to 0, y_3(t) \to 0$$

Thus we still have

$$\|x(t)\| < \infty, x(t) \to x_1(t), y(t) \to y_1(t) \to y_r(t)$$

That completes the proof. □

Of course, $d(t)$ may have some effects on the performances of the weighting algorithm and the closed-loop system. If only Eq. (6.12) can be satisfied, then $d(t)$ does not make difference to the stability and tracking performance of the WMMAC system.

In Theorem 6.2 we actually deal with an ideal modeling situation, next we consider the non-ideal modeling situation, i.e., the true model of the plant is not included in the model set, but there is a closet one with bounded (or vanishing) modeling errors, i.e., $e_l(t) \triangleq y_p(t) - y_{ml}(t)$, $e_{xl}(t) \triangleq x_p(t) - x_{ml}(t)$ are bounded or further $\|e_l(t)\| \to 0$, $\|e_{xl}(t)\| \to 0$, Then we have the following results.

Corollary 6.1 *The proposed WMMAC system for plant described in Eq. (6.1) with weighting algorithm (6.5)–(6.11) is stable, if only the following conditions are satisfied:*

(1) There is a closest local model to the plant, denoted as $M_l \in \mathbb{M}$, with bounded modeling errors $e_l(t)$, $e_{xl}(t)$, such that condition (6.12) is satisfied;
(2) Each local controller C_i is stable itself, C_i and M_i constitute an internal stable closed-loop system whose output is tracking the reference signal $y_r(t)$;
(3) Disturbance $d(t)$ is bounded.

Corollary 6.2 *The proposed WMMAC system for plant described in Eq. (6.1) with weighting algorithm (6.5)–(6.11) is stable and tracking the reference signal, if only the following conditions are satisfied:*

(1) There is a closest local model to the plant, denoted as $M_l \in \mathbb{M}$, with $\|e_l(t)\| \to 0$, $\|e_{xl}(t)\| \to 0$, such that condition (6.12) is satisfied;
(2) Each local controller C_i is stable itself, C_i and M_i constitute an internal stable closed-loop system whose output is tracking the reference signal $y_r(t)$;
(3) Disturbance $d(t)$ is bounded and tends to zero.

The proofs of Corollaries 6.1 and 6.2 are similar to that of Theorem 6.2, details are omitted here to save space.

Corollaries 6.1 and 6.2 mean that although only limited discrete parameter uncertainties are considered, the proposed WMMAC scheme can also be applied to other uncertain situations, such as the uncertain parameter is a continuum, or even with unstructured uncertainties.

6.4 Simulation Results

This section give some simulation results to verify the effectiveness of the theoretical analysis of the proposed WMMAC scheme. We consider the well-known benchmark problem, i.e., the two-cart mass-spring-damper (MSD) system [1], which was shown in Fig. 6.7. The performances of the proposed WMMAC scheme are compared with that of the corresponding RMMAC system, and the non-adaptive robust control system, respectively. In Fig. 6.7, the parameters $m_1 = m_2 = 1$ and $b_1 = b_2 = 0.1$ are known, while k_1 is known to belong to $\Omega = k_1 : 0.25 \leq k_1 \leq 1.45$. The disturbance

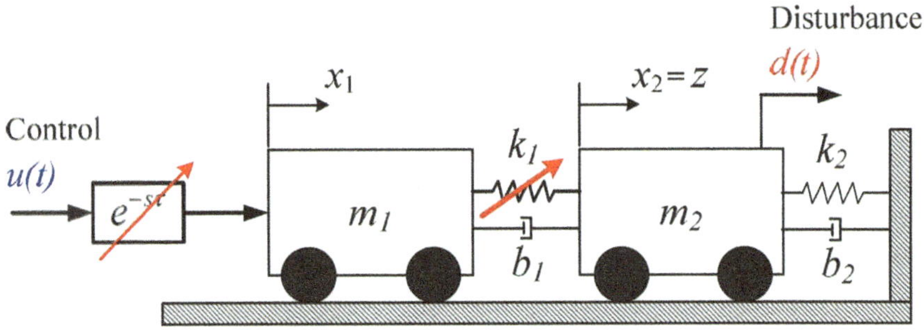

Fig. 6.7 The two-cart mass-spring-damper (MSD) system

$d(t)$ is a low-frequency, stationary stochastic process that acts on m_2 and is generated by

$$d(s) = \frac{a}{s+a}\omega(s)$$

where $a = 0.1$ is the disturbance bandwidth and the white gaussian process $\omega(t)$ has zero mean and unit intensity. The only measurement is m_2's displacement plus additive white gaussian noise, i.e., $y(t) = x_2(t) + v(t)$, where $Ev(t) = 0$ and $Ev(t)v(\tau) = 10^{-6}\delta(t-\tau)$. The control is applied to m_1 through a control channel with an maximum time delay of 0.05 s, i.e.

$$\tau \le 0.05 \quad \text{s}$$

The state-space description of the two-cart MSD system is given by

$$\begin{cases} \dot{x}(t) = Ax(t) + Bu(t) + L\omega(t) \\ y(t) = Dx(t) + v(t) \\ x(0) = 0 \end{cases} \tag{6.21}$$

where the state vector is $x(t) = [x_1(t) \quad x_2(t) \quad \dot{x}_1(t) \quad \dot{x}_2(t) \quad d(t)]^T$ and

$$A = \begin{bmatrix} 0 & 0 & 1 & 0 & 0 \\ 0 & 0 & 0 & 1 & 0 \\ -\frac{k_1}{m_1} & \frac{k_1}{m_1} & -\frac{b_1}{m_1} & \frac{b_1}{m_1} & 0 \\ \frac{k_1}{m_2} & -\frac{k_1+k_2}{m_2} & \frac{b_1}{m_2} & -\frac{b_1+b_2}{m_2} & \frac{1}{m_2} \\ 0 & 0 & 0 & 0 & -a \end{bmatrix} \tag{6.22}$$

$$B = \left[0\ 0\ \tfrac{1}{m_1}\ 0\ 0\right]^T \tag{6.23}$$

$$L = \left[0\ 0\ 0\ 0\ a\right]^T \tag{6.24}$$

$$C = \begin{bmatrix} 0 & 1 & 0 & 0 & 0 \end{bmatrix}^T \tag{6.25}$$

The control objective is to keep variable $z(t) = x_2(t)$ as small as possible, i.e., the disturbance attenuation. Then we need to design suitable controller $u(t)$ so that the effects of disturbance and noise on $z(t)$ are attenuated as much as possible.

The uncertainties of k_1 are divided into three subdivisions, accordingly the following three models are used to construct the 'local' controllers by mixed-μ-synthesis method, for more details, see references [1, 15, 16].

$$\begin{aligned} M_1 : &\quad 0.25 \le k_1 \le 0.35 \Rightarrow k_1 = 0.3 \\ M_2 : &\quad 0.35 \le k_1 \le 0.65 \Rightarrow k_1 = 0.5 \\ M_3 : &\quad 0.65 \le k_1 \le 1.45 \Rightarrow k_1 = 1.05 \end{aligned}$$

Controllers C_1, C_2, and C_3 are designed according to M_1, M_2, and M_3, respectively.

The following two cases are considered in simulation.

Case 1: k_1 is fixed, i.e. $k_1 = 0.6$. The performance of weighting algorithm (6.5)–(6.11) is shown in Fig. 6.8. The performance of weighting algorithm of RMMAC is shown in Fig. 6.9. The control (disturbances attenuation) performances of WMMAC, RMMAC, and the non-adaptive robust control, are shown in Fig. 6.10. The performances og WMMAC and RMMAC are almost the same, and significantly better than the non-adaptive robust control system.

Case 2: k_1 is jumping from $k_1 - 0.6$ to $k_1 = 1.45$ at $t = 40\,s$.. The control (disturbances attenuation) performances of WMMAC, RMMAC, and the non-adaptive robust control, are shown in Fig. 6.13. The corresponding weight signals of weighting algorithm (6.50)–(6.11) and the weighting algorithm of RMMAC are shown in Figs. 6.11, and 6.12, respectively. The WMMAC performs a little better than RMMAC, and they both perform significantly better than the non-adaptive robust control. It should also be pointed out that, in Case 2, there are two methods to avoid the zero lock-in condition of p_i from occurring. One is to set lower bounds for

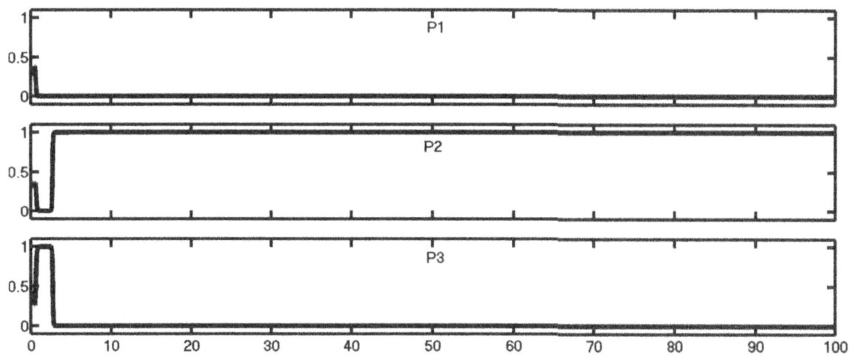

Fig. 6.8 Performance of weighting algorithm of WMMAC in CASE 1

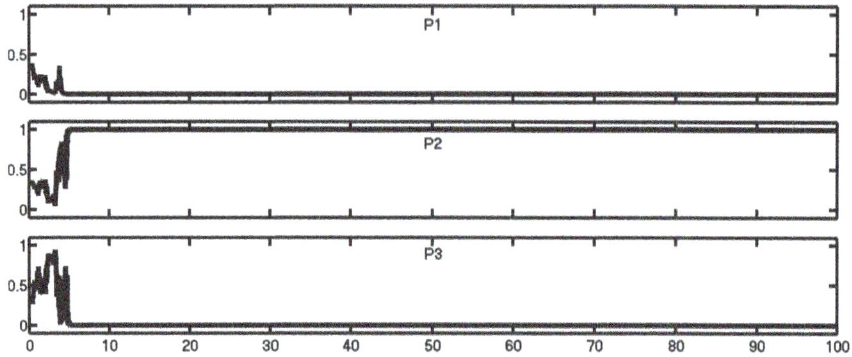

Fig. 6.9 Performance of weighting algorithm of RMMAC in CASE 1

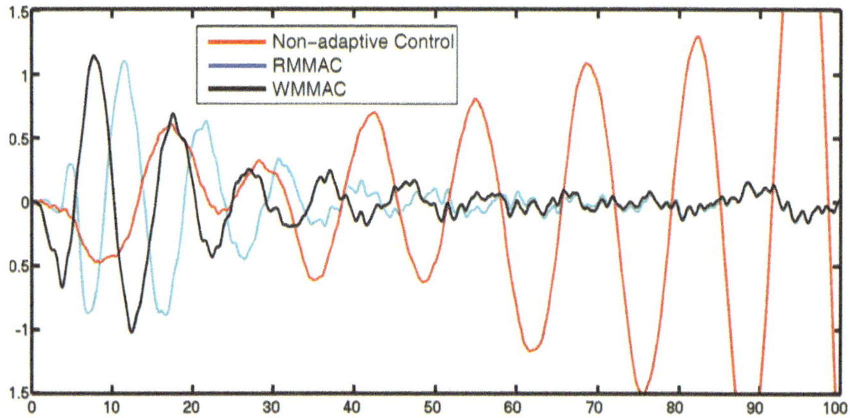

Fig. 6.10 Performances of WMMAC, RMMAC, and non-adaptive control systems in CASE 1

weights $p_i, i = 1, 2, \ldots, m$ as in reference [17]; The other is to reset each weight according to Eq. (6.5) when the parameter jump is detected.

In summary, all these simulation results, as shown in Figs. 6.8, 6.9, 6.10, 6.11, 6.12, 6.13, respectively, verified the effectiveness of the WMMAC scheme including the weighting algorithm.

6.5 Conclusion

The stability of WMMAC (including RMMAC) of continuous-time plant has been proved under some reasonable assumptions, i.e., model set covers the uncertainties of the plant to be controlled, and in the model set, there is a closest model of the plant, which generates the minimum model output error; each local controller is designed to

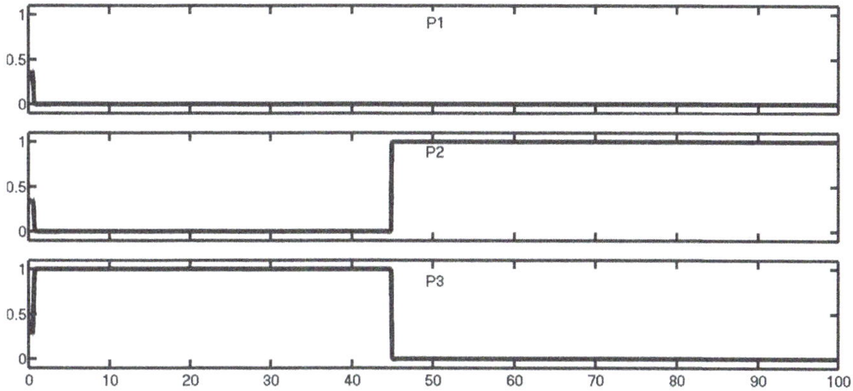

Fig. 6.11 Performance of weighting algorithm of WMMAC in CASE 2

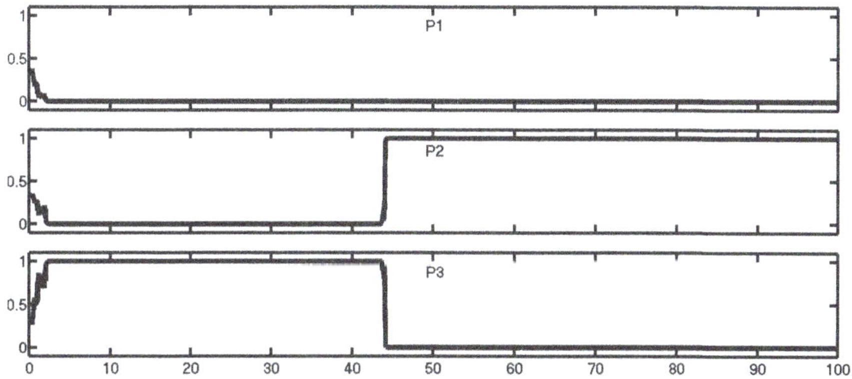

Fig. 6.12 Performance of weighting algorithm of RMMAC in CASE 2

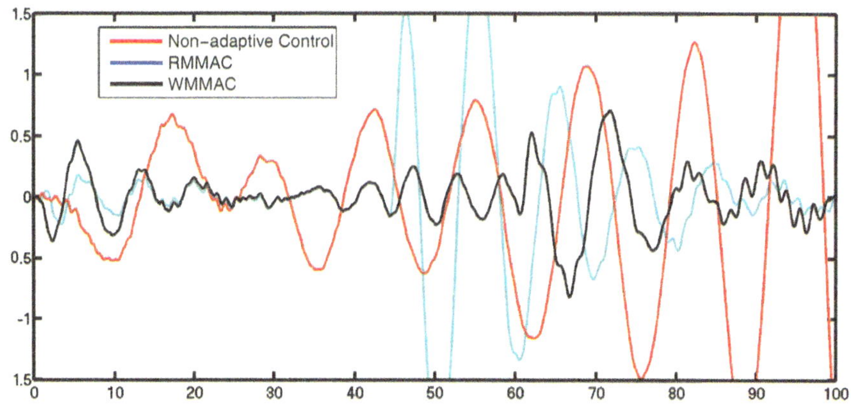

Fig. 6.13 Performances of WMMAC, RMMAC, and non-adaptive control systems in CASE 2

stabilize its corresponding local model and tracking the reference signal. It should be noticed that this chapter didn't pay much attention to the modeling problem (model set construction), which has been discussed thoroughly in references [17–20]. And in the authors' opinion, the modeling problem is more related to the performance requirements of the specific control engineering.

We discussed the stability issue of WMMAC in three chapters, but we didn't consider another important type of MMAC, i.e., switching MMAC. Actually, switching MMAC could be regarded as a special case of WMMAC. Some related research works will be published soon in the near future.

References

1. S. Fekri, M. Athans, A. Pascoal, Issues, progress and new results in robust adaptive control. Int. J. Adapt. Control Signal Process. **20**(10), 519–579 (2006)
2. D.T. Magill, Optimal adaptive estimation of sampled stochastic processes. IEEE Trans. Autom. Control **10**, 434–439 (1965)
3. D.G. Lainiotis, Partitioning: a unifying framework for adaptive systems-I: estimation; II: control. Proc. IEEE **64**, 1126–1143 and 1182–1197 (1976)
4. M. Athans et al., The stochastic control of the F-8C aircraft using a multiple model adaptive control (MMAC) method-Part I: equilibrium flight. IEEE Trans. Autom. Control **22**, 768–780 (1977)
5. A. Badr, Z. Binder, D. Rey, Weighted multi-model control. Int. J. Syst. Sci. **23**(1), 145–149 (1992)
6. M. Kuipers, P. Ioannou, Practical robust adaptive control: benchmark example, in *Proceedings of American Control Conference* (Seattle, Washington, USA, 2008), pp. 5168–5173
7. M. Kuipers, P. Ioannou, Multiple model adaptive control with mixing. IEEE Trans. Autom. Control **55**(8), 1822–1836 (2010)
8. S. Baldi, P.A. Ioannou, E. Mosca, Multiple model adaptive mixing control: the discrete-time case. IEEE Trans. Autom. Control **57**(4), 1040–1045 (2012)
9. N. Sadati, G.A. Dumont, H.R. Feyz Mahdavian, Robust multiple model adaptive control using fuzzy fusion, in *42nd South Eastern Symposium on System Theory* (Tyler, TX, USA, 2010)
10. Z. Han, K.S. Narendra, New concepts in adaptive control using multiple models. IEEE Trans. Autom. Control **57**(1), 78–89 (2012)
11. K.S. Narendra, A.M. Annaswamy, *Stable Adaptive Systems* (Prentice-Hall, Englewood Cliffs, NJ, 1989)
12. W. Zhang, Stable weighted multiple model adaptive control: discrete-time stochastic plant. Int. J. Adapt. Control Signal Process. **27**(7), 562–581 (2013)
13. W. Zhang, Further results on stable weighted multiple model adaptive control: discrete-time stochastic plant. Int. J. Adapt. Control Signal Process. **29**(12), 1497–1514 (2015)
14. W. Zhang, On the stability and convergence of self-tuning control-virtual equivalent system approach. Int. J. Control **83**(5), 879–896 (2010)
15. S. Fekri, M. Athans, A. Pascoal, RMMAC: a novel robust adaptive control scheme-Part I: architecture, in *Proceedings of the IEEE Conference on Design and Control* (Paradise Island, Bahamas, 2004), pp. 1134–1139
16. S. Fekri, M. Athans, A. Pascoal, RMMAC: a novel robust adaptive control scheme-part II: performance evaluation, in *Proceedings of the IEEE Conference on Design and Control* (Paradise Island, Bahamas, 2004), pp. 1140–1145
17. J.A. Custafson, P.S. Maybeck, Flexible spacestructure control via moving-bank multiple model algorithms. IEEE Trans. Aerospace Electr. Syst. **30**(3), 750–757 (1994)

18. H. Mahdianfar, S. Ozgoli, H.R. Momeni, Robust multiple model adaptive control: modified using ν-gap metric. Int. J. Robust Nonlinear Control **21**(18), 2027–2063 (2011)
19. M. Huang, X. Wang, Z. Wang, Multiple model self-tuning control for a class of nonlinear systems. Int. J. Control **88**(10), 1984–1994 (2015)
20. J. Bernat, S. Stepien, Multi-modelling as new estimation schema for high-gain observers. Int. J. Control **88**(6), 1209–1222 (2015)

Chapter 7
Stability of Continuous-Time T-S Model Based Fuzzy Control Systems

Abstract Fuzzy control is an important branch of control theory and engineering. This chapter is aimed to provides a unified stability analysis framework based on VES concept and methodology for a class of continuous-time T-S model based fuzzy control (TSFC) systems, which consist of parallel distributed compensation (PDC) strategy and the real plant instead of the T-S fuzzy model. Plant to be controlled includes linear time-invariant, linear parameter jump, or nonlinear time-varying systems. As an alternative to Lyapunov function based stability analysis approach, VES approach is adopted as a unified stability analysis tool and methodology.

7.1 Introduction

Although many fuzzy control strategies have been widely used in different kinds of control engineering applications [1], there is still a long way to go for reaching a unified stability theory for this important control branch. TSFC [2] shed some lights on the issue, which provides an alternative way to the modeling and control of a wide class of complex nonlinear systems. Also TSFC significantly reduces the number of fuzzy rules in describing complex nonlinear systems with higher order [3]. More importantly, TSFC is a basis towards a systematic approach of analyzing and designing a fuzzy control system. Many theoretical research results, such as, function approximation capability of T-S model, stability analysis approach, and controller synthesis strategy, have been developed [4, 5]. As well known, T-S model can be used to approximate any smooth nonlinear functions with arbitrary required accuracy [6–12]. These results laid a sound theoretical foundation for using T-S fuzzy model to describe complex nonlinear systems approximately.

It is obvious to see that TSFC actually paves a bridge between classical control theory and fuzzy control system analysis. With the help of TSFC, it is possible to analyze the stability of fuzzy control systems based on classical control theory. It should also be pointed out that TSFC can be combined with any existing control strategies [44–46].

To date, many methodologies have been suggested for the stability analysis of TSFC systems, for example, common quadratic Lyapunov function method [12],

W. Zhang and Q. Li, *Virtual Equivalent System Approach for Stability Analysis of Model-based Control Systems*, https://doi.org/10.1007/978-981-15-5538-1_7

piecewise quadratic Lyapunov function method [13–15], fuzzy Lyapunov function method [16–20], etc.

Unfortunately, owing to the lack of suitable analysis tools, there are different kinds of drawbacks (conservativeness) in the stability analysis of TSFC with above-mentioned methods. In addition, fuzzy rules and fuzzy membership functions have not been extensively explored in these methods [47]. Also, there are many other problems that remain unsolved in TSFC systems, for more details, see reference [5].

Motivated by these observations, VES approach [21] was introduced as a unified framework to address the stability issue of TSFC systems. In fact, VES is an artificial or man-made system that is equivalent to the original control system in the input-output sense. With the help of VES concept and methodology, original nonlinear system problem can be readily converted into a linear system problem with a nonlinear compensation signal under some certain conditions (for details, see related contents in later sections). In other words, the nonlinearity in structure of the original control system has been transferred to the compensation signal in the VES system.

In the beginning, VES concept and corresponding analysis methodology has been put forward for understanding (or judging) the stability and convergence performances of a general STC system in a unified manner. VES approach is independent of specific self-tuning control strategy, specific parameter estimates algorithm, and specific plant (deterministic or stochastic) to be controlled. Later on, VES approach has also been adopted to analyze the stability of WMMAC systems [22, 23]. In the authors' opinion, VES approach is suitable for stability analysis of any model based control systems.

It should be pointed out that some related preliminary results have been published in journals and conference proceedings, see for example, references [24, 25, 49], in which discrete-time TSFC systems are also considered.

7.2 Description of a TSFC System

There are mainly two types of TSFC systems, one is the smooth fuzzy control, i.e., parallel distributed compensation (PDC), and the other is the switching control, i.e., the local compensation scheme. This chapter will be focused mainly on the PDC scheme, which is very similar to WMMAC in system structure. To be specific, the normalized membership function in TSFC systems plays the same role as the weighting algorithm in WMMAC systems.

T-S fuzzy models, also known as fuzzy dynamic models, have been widely used to represent some complex systems with both fuzzy inference rules and local analytic linear dynamic models as follows:

$$R^l: \quad \text{IF } z_1 \text{ is } F_1^l, z_2 \text{ is } F_2^l \text{ and} \ldots z_v \text{ is } F_v^l$$

Then

$$\begin{aligned} \dot{x}(t) &= A_l x(t) + B_l u(t) \\ y(t) &= C_l x(t) \\ l &\in L := 1, 2, \ldots, m \end{aligned} \tag{7.1}$$

Obviously, state-space model (7.1) can also be equivalently transformed into polynomial (operator) model, i. e.,

$$\begin{aligned} Y(s) &= G_l(s)U(s) \\ l &\in L := 1, 2, \ldots, m \end{aligned} \tag{7.2}$$

where, R^l denotes the lth fuzzy rule, m the number of fuzzy rules, $F_j^l (j = 1, \ldots, v)$ the fuzzy sets, $x(t) \in \mathbb{R}^n$ the state vector, $u(t) \in \mathbb{R}^1$ the input signal of the plant, $y(t) \in \mathbb{R}^1$ the output signal of the plant, $x(0) = 0$, $u(0) = 0$, $y(0) = 0$, $X(s) = \mathcal{L}[x(t)]$, $U(s) = \mathcal{L}[u(t)]$, $Y(s) = \mathcal{L}[y(t)]$, $\mathcal{L}[\cdot]$ stands for the Laplace transform; (A_l, B_l, C_l) the matrices of the lth local model, and $z(t) := [z_1, z_2, \ldots, z_v]$ some measurable variables of the system, for example, the state variables.

For simplicity, denote the local model set in (7.1) and (7.2) as $M_l, l \in L := 1, 2, \ldots, m$.

Local model set should be constructed appropriately to cover the uncertainties of the controlled plant. By 'cover' we mean that one of the following situations (for more detailed classification of the uncertainties of plant, please see the stability analysis in later sections):

(1) Controlled plant is an unknown LTI system. The true model of plant is included in the local model set;
(2) Controlled plant is an unknown LTI system. The true model of plant is not included in the local model set, but there is a closest one to the plant, with approximation error $e(t) = o(\beta + \|x(t)\|)$, or $\|e(t)\| \leq M < \infty$;
(3) Controlled plant is an unknown LTI system. The true model of plant is not included in the local model set, but there are some local models which are equally close to the plant;
(4) Controlled plant is a linear parameter jump or nonlinear time-varying system, but it can be approximated by multiple linear local models with small enough approximation error, i. e., $e(t) = 0$, or $e(t) = o(\beta + \|x(t)\|)$, or $\|e(t)\| \leq M < \infty$.

The definition of the notation Little-Oh 'o' can be found in Appendix D. It should be noticed that different system models (state-space model or transfer function model) will be used in different situations for convenience.

Throughout the paper, the control objective is the stability of the closed-loop TSFC system, and the tracking performance of system output, i.e., $y(t) \to y_r(t)$, $y_r(t)$ is a bounded reference signal. Of course, each local controller C_l is designed according to corresponding local model $M_l, l \in L := 1, 2, \ldots, m$.

Without loss of generality, we consider the local controller C_l in the form of state feedback

$$u_l(t) = K_l x(t) + H_l y_r(t) \tag{7.3}$$

or taking the form of output feedback

$$u_l(s) = C_l(s)[y_r(s) - y(s)] \tag{7.4}$$

Generally speaking, as design objective, C_l needs to stabilizes $M_l, l \in L := 1, 2, \ldots, m$, and the output of the resulting closed-loop control system tracks the bounded reference signal $y_r(t)$.

Furthermore the global control law is obtained by

$$u(t) = \sum_{l=1}^{m} \mu_l(t) u_l(t) \tag{7.5}$$

where $\mu_l, l = 1, \ldots m$ is the normalized membership function satisfying the following equations

$$\mu_l = \frac{\xi_l(z)}{\sum_{j=1}^{m} \xi_j(z)} \tag{7.6}$$

$$\xi_l(z) = \prod_{i=1}^{v} F_i^l(z_i) \tag{7.7}$$

$$\mu_l \geq 0, \quad \sum_{l=1}^{m} \mu_l = 1 \tag{7.8}$$

and $F_i^l(z_i)$ is the grade of membership of z_i in the fuzzy set F_i^l.

The block diagrams of TSFC systems in state feedback form and output feedback form, are shown in Fig. 7.1a, b, respectively, where P represents the uncertain plant to be controlled.

From Fig. 7.1a, b, it is not difficult to see that TSFC is rather similar to weighted multiple model adaptive control (WMMAC) [26–29] . Actually, there are also sim-

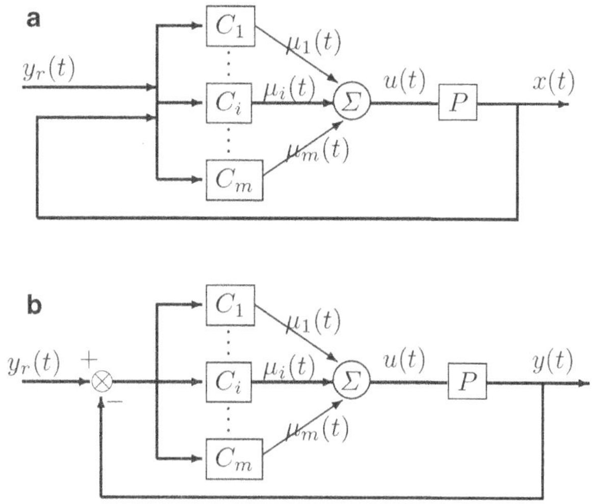

Fig. 7.1 **a** Block diagram of TSFC of LTI plant in state feedback form. **b** Block diagram of TSFC of LTI plant in output feedback form

ilarities pretty much between gain scheduling control and TSFC. Gain scheduling control systems are widely adopted to control uncertain non-linear systems [30–34].

In Fig. 7.1a, b, $\mu_i, i = 1, \ldots m$ actually represent the model identification of plant P. For an LTI plant, the situation of $\mu_i, i = 1, \ldots m$ can be divided into two cases, i.e., 'ideal identification' and 'non-ideal identification'. By 'ideal identification' we mean that $\mu_l \to 1, \mu_i \to 0, i \neq l$, otherwise we refer to 'non-ideal identification'. In the following contents, we will deal with these two cases separately.

7.3 Stability Analysis of TSFC for LTI Plant: Ideal Identification

As mentioned before, 'ideal identification' means that $\mu_l \to 1, \mu_i \to 0, i \neq l$. In this section, two situations will be considered. In situation 1, the true model of plant is included in the local model set; in situation 2: the true model of plant is not included in the local model set, but there is a closest one to the plant.

7.3.1 VES Constrction of TSFC Systems

First we consider situation 1. Suppose M_l is the true model of plant P, and that $\mu_l \to 1, \mu_i \to 0, i \neq l$. In this situation, the VES for TSFC system (as shown in Fig. 7.11a) can be constructed as shown in Fig. 7.2, where $u_l(t)$ is the output of local controller C_l, $u(t)$ is the same as in Fig. 7.11a, $\Delta u(t) = u(t) - u_l(t)$.

Fig. 7.2 VES I of Fig. 7.1a

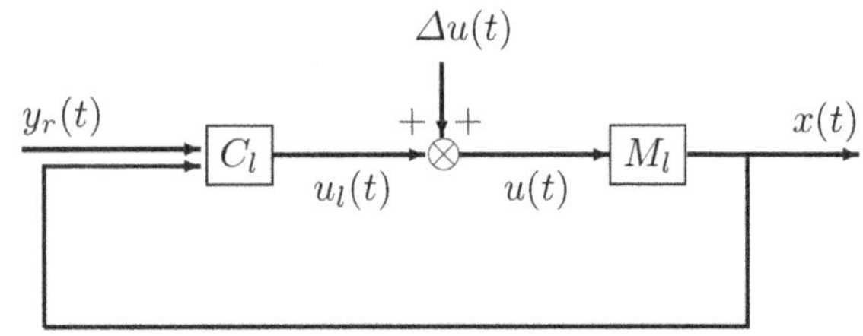

Fig. 7.3 VES II of Fig. 7.1a

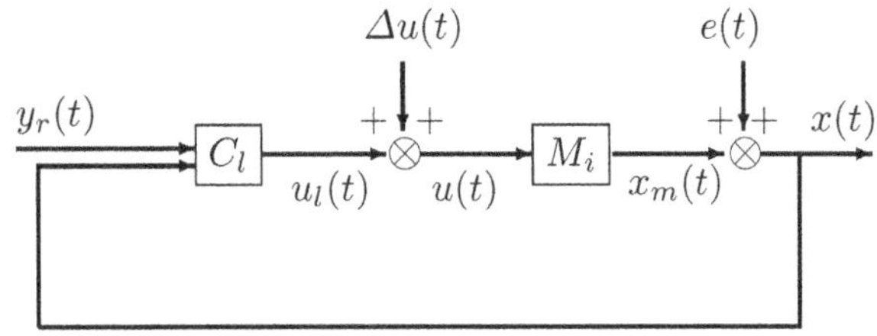

Similarly, we can also construct the VES II as shown in Fig. 7.3 for situation 2 (the true model of plant is not included in the model set), where M_l is the closest model in local model set to plant P, $x_m(t)$ is the state vector generated by M_l, and $e(t) = x(t) - x_m(t)$ is the corresponding state approximation error.

Next, we give stability criteria on TSFC systems according to VES I, and VES II, respectively. In detail, Theorem 7.1 will be developed according to VES I, Theorems 7.2 and 7.3 will be developed according to VES II.

7.3.2 Main Result

This section includes the main results on the stability of TSFC systems based on VES concept and methodology. To be specific, Theorem 7.1 gives the stability criterion for situation 1 (the true model of plant is included in the local model set); Theorems 7.2 and 7.3 give the criteria for situation 2 (the true model of plant is not included in the local model set).

Theorem 7.1 *The TSFC system as shown in Fig. 7.1a is stable and tracking the reference signal if only the following conditions are satisfied:*

(1) Local model set in fuzzy rules covers the uncertainties of the plant, i.e., the true model of plant, say M_l, is included in the local model set;
(2) Fuzzy rules and the membership functions "identify" the true model of the plant, in other words, the normalized membership function converge correctly, i.e., $\mu_l \to 1, \mu_i \to 0, i \neq l$;

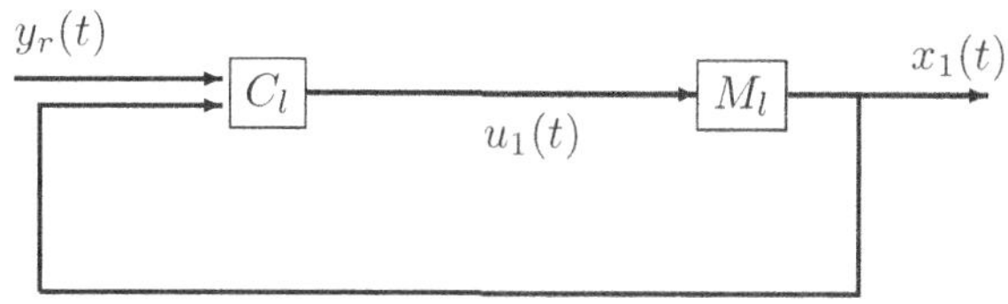

Fig. 7.4 Decomposed subsystem I of VES I (Fig. 7.2)

(3) Each local controller together with its corresponding local model formulates a stable and tracking closed-loop control system.

Proof From Conditions (1) and (2) we know that Figs. 7.1a, and 7.2 are equivalent in the input-output sense. Further, based on Condition (2), we obtain the property of $\Delta u(t)$ as follows.

$$\begin{aligned}
\Delta u(t) &= u(t) - u_l(t) \\
&= \sum_{i=1}^{m} \mu_i [K_i x(t) + H_i y_r(t)] - [K_l x(t) + H_l y_r(t)] \\
&= \mu_1 [K_1 x(t) + H_1 y_r(t)] + \cdots + \\
&\quad \mu_l [K_l x(t) + H_l y_r(t)] + \cdots + \\
&\quad \mu_m [K_m x(t) + H_m y_r(t)] - \\
&\quad K_l x(t) - H_l y_r(t) \\
&= o(\|x(t)\| + \alpha)
\end{aligned} \tag{7.9}$$

where $\alpha > 0$ is the upper bound of reference signal $y_r(t)$, i.e., $\|y_r(t)\| \leq \alpha$.

Next, noticing that the VES I (Fig. 7.2) represents a linear system in structure. It can be decomposed into two subsystems, as shown in Figs. 7.4 and 7.5, respectively.

Considering that all these systems, i.e., Figs. 7.2, 7.4, and 7.5 have initial values of zero, then according to linear superposition principle, we have

$$x(0) = x_1(0) + x_2(0) \tag{7.10}$$

$$u(0) = u_1(0) + u_2(0) \tag{7.11}$$

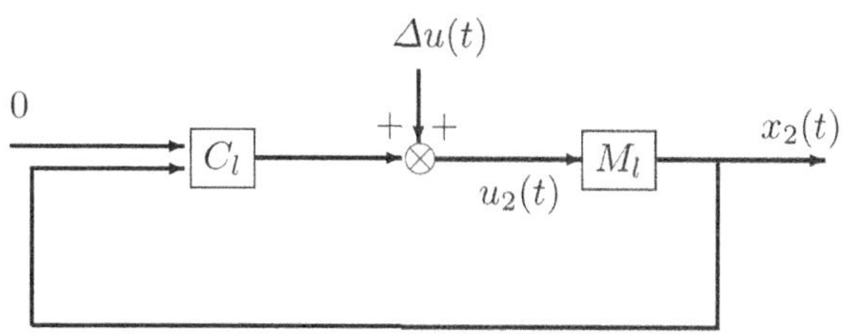

Fig. 7.5 Decomposed subsystem II of VES I (Fig. 7.2)

$$y(0) = y_1(0) + y_2(0) \tag{7.12}$$

$$x(t) = x_1(t) + x_2(t) \tag{7.13}$$

$$u(t) = u_1(t) + u_2(t) \tag{7.14}$$

$$y(t) = y_1(t) + y_2(t) \tag{7.15}$$

By Condition (3) we know that subsystems as shown in Figs. 7.4 and 7.5 are both stable and tracking (the reference signal).

Then we have in Fig. 7.4

$$\|x_1(t)\| < \infty \tag{7.16}$$

$$y_1(t) \to y_r(t) \tag{7.17}$$

In Fig. 7.5, we have (for details, see reference [35])

$$x_2(t) = O(\|\Delta u(t)\|) \tag{7.18}$$

The definition of Big-Oh 'O' is given in Appendix D.

Putting Eqs. (7.18) and (7.9) together, we get

$$x_2(t) = o(\|x(t)\| + \alpha) \tag{7.19}$$

Equation (7.19) together with Eq. (7.13) yields

$$\begin{aligned} 0 \le \|x(t)\| &\le \|x_1(t)\| + \|x_2(t)\| \\ &= \|x_1(t)\| + o(\|x(t)\| + \alpha) \end{aligned} \tag{7.20}$$

Next we prove the final conclusion of Theorem 7.1 by principle of reduction to absurdity.

Suppose $\|x(t)\|$ is unbounded, then there must exists a subseries $\|x(t_k)\|$, $\|x(t_k)\| \to \infty$, thus from (7.20) we have

$$0 \le \|x(t_k)\| \le \|x_1(t_k)\| + o(\|x(t_k)\| + \alpha) \tag{7.21}$$

Dividing both sides of Eq. (7.21) with $(\|x(t_k)\| + \alpha)$, we have

$$0 \le \frac{\|x(t_k)\|}{\|x(t_k)\| + \alpha} \le \frac{\|x_1(t_k)\|}{\|x(t_k)\| + \alpha} + o(1) \tag{7.22}$$

Considering $\|x_1(t_k)\|$ is bounded, and $\|x(t_k)\| \to \infty$, then we have

$$\frac{\|x_1(t_k)\|}{\|x(t_k)\| + \alpha} + o(1) \to 0 \tag{7.23}$$

By Squeezing Principle, from Eqs. (7.22) and (7.23) we obtain

$$\frac{\|x(t_k)\|}{\|x(t_k)\| + \alpha} \to 0 \tag{7.24}$$

Obviously, (7.24) is contradictory to the fact $\|x(t_k)\| \to \infty$.

Thus we conclude the former assumption, i.e., $\|x(t)\|$ is unbounded, can not hold. Then we have

$$\|x(t)\| < \infty \tag{7.25}$$

Further, taking Eqs. (7.19), (7.13), (7.15), and (7.17) into account, we obtain the following results

$$x_2(t) \to 0 \tag{7.26}$$

$$y_2(t) = C_l x_2(t) \to 0 \tag{7.27}$$

$$x(t) \to x_1(t) \tag{7.28}$$

$$y(t) \to y_1(t) \to y_r(t) \tag{7.29}$$

i.e., VES I (Fig. 7.2) is stable and tracking (the reference signal), so is the TSFC system for LTI plant (as shown in Fig. 7.1a).

That completes the proof.

For situation 2 (the true model of plant is not included in the local model set), we have the following result.

Theorem 7.2 *The TSFC system for as shown in Fig. 7.1a is stable and tracking the reference signal, if only the following conditions are satisfied:*

(1) M_l is the closest model to the plant, with state approximation error

$$e(t) = x(t) - x_m(t) = o(\beta + \|x(t)\|)$$

where $\beta > 0$ is a constant;

(2) Fuzzy rules and the membership functions "identify" the closest model to the plant, in other words, the normalized membership function converge correctly, i.e., $\mu_l \to 1, \mu_i \to 0, i \neq l$;

(3) Each local controller together with its corresponding local model formulates a stable and tracking closed-loop control system.

Proof Similar to the proof of Theorem 7.1, VES II (Fig. 7.3) can be decomposed into three subsystems, i.e, Figs. 7.6, 7.7, and 7.8, respectively.

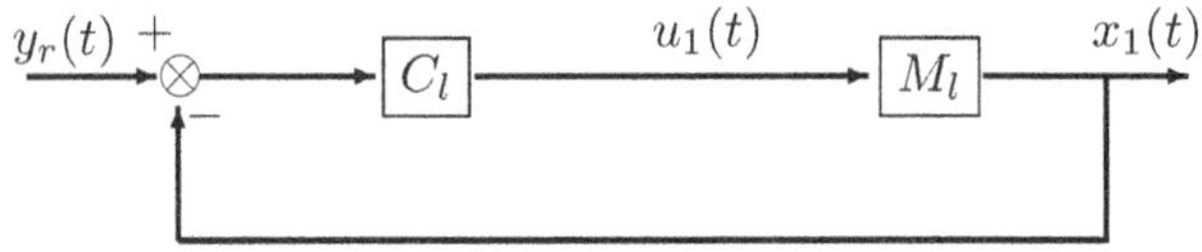

Fig. 7.6 Decomposed subsystem 1 of VES II (Fig. 7.3)

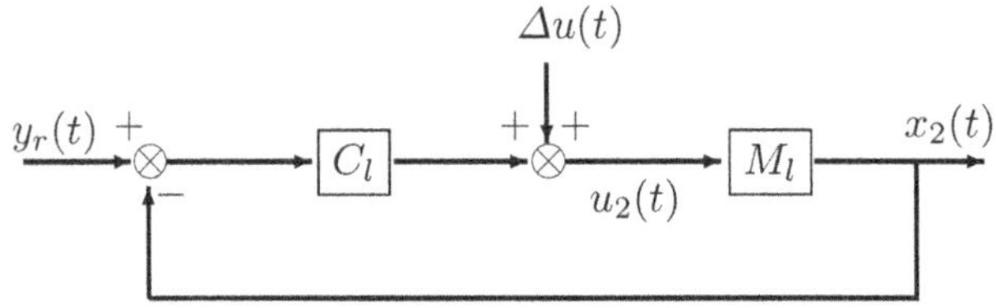

Fig. 7.7 Decomposed subsystem 2 of VES II (Fig. 7.3)

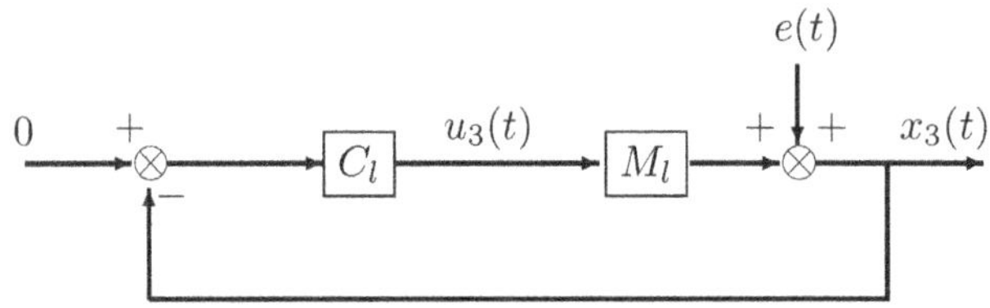

Fig. 7.8 Decomposed subsystem 3 of the VES II (Fig. 7.3)

Then we have

$$x(0) = x_1(0) + x_2(0) + x_3(0) \tag{7.30}$$

$$u(0) = u_1(0) + u_2(0) + u_3(0) \tag{7.31}$$

$$y(0) = y_1(0) + y_2(0) + y_3(0) \tag{7.32}$$

$$x(t) = x_1(t) + x_2(t) + x_3(t) \tag{7.33}$$

$$u(t) = u_1(t) + u_2(t) + u_3(t) \tag{7.34}$$

$$y(t) = y_1(t) + y_2(t) + y_3(t) \tag{7.35}$$

The rest procedures of the proof follows along the similar lines as in the proof of Theorem 7.1, and is thus omitted here to save space.

To further relax the conditions, we have the following theorem regarding the stability (boundedness) of the TSFC system for an LTI plant.

Theorem 7.3 *The TSFC system for an LTI plant as shown in Fig. 7.1a is stable (boundedness), if only the following conditions are satisfied:*

(1) M_l is the closest model to the plant, with bounded state approximation error $e(t) = x(t) - x_m(t)$, i.e., $\|e(t)\| \leq \gamma < \infty$;

(2) Fuzzy rules and the membership functions correctly "identify" the closest model to the plant, in other words, the normalized membership function converge correctly, i.e., $\mu_l \to 1, \mu_i \to 0, i \neq l$;

(3) Each local controller together with its corresponding local model formulates a stable and tracking closed-loop control system.

Proof The only difference between Theorems 7.3 and 7.2 is the property of $e(t)$, i.e, in Theorem 7.2 we have $e(t) = x(t) - x_m(t) = o(\beta + \|x(t)\|)$; whereas in Theorem 7.3, we have $\|e(t)\| \leq \gamma < \infty$.

Then following the similar procedures of the proof of Theorems 7.2, 7.3 can be readily proved, details are omitted here to save some space.

7.4 Stability Analysis for LTI Plant: Non-ideal Identification

In this section we will consider two situations, i.e., situation 1: $\mu_i, i = 1, \dots m$ does not converge; situation 2: $\mu_i, i = 1, \dots m$ converges, but not converge to '0' or '1'. Owing to these two special situations and the reason of analysis tool, we will not differ between these two situations, for more details, see later contents of analysis.

7.4.1 VES of TSFC Systems

It would be reasonable, for these two situations, to suppose that the true model of plant P is not included in the local model set, but each local model is 'close' to the plant. Thus, the VES for TSFC (Fig. 7.1b) can be constructed as shown in Fig. 7.9, where $\overline{C} = \sum_{i=1}^{m} \mu_i C_i$.

As stated before, such situations actually mean that all the local models are close to the true model of the plant, thus they are undistinguishable from each other. To be specific, let

$$\begin{cases} P = \frac{N(s)}{D(s)} \\ M_l = \frac{N_{ml}(s)}{D_{ml}(s)} \\ C_l = \frac{N_{cl}(s)}{D_{cl}(s)} \end{cases} \quad l \in L \tag{7.36}$$

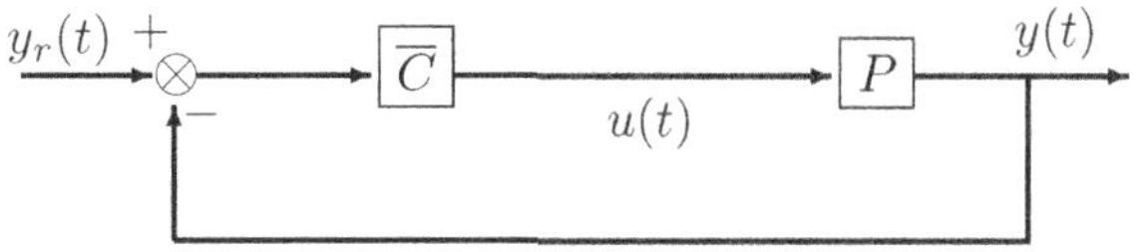

Fig. 7.9 VES III of Fig. 7.1b

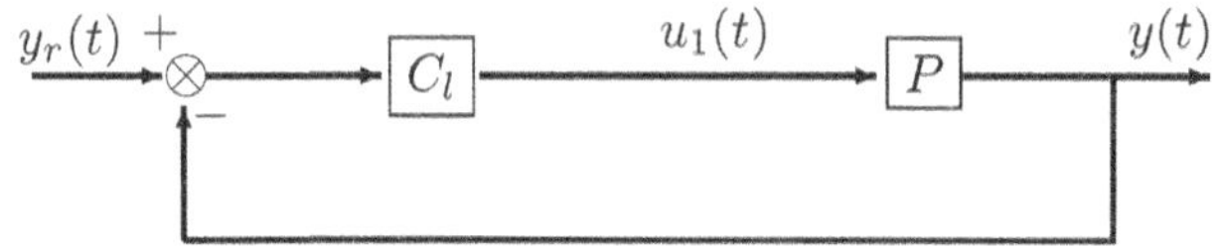

Fig. 7.10 Block diagram for explaining some notations in Sect. 7.4

Then we have

$$C_l M_l = \frac{N_{cl}(s)}{D_{cl}(s)} \frac{N_{ml}(s)}{D_{ml}(s)} \tag{7.37}$$

$$C_l P = \frac{N_{cl}(s)}{D_{cl}(s)} \frac{N(s)}{D(s)} \tag{7.38}$$

By the word 'close' we mean that if C_l stabilizes $M_l, l \in L$ both in open- and closed-loop, i.e., $D_{cl}(s)D_{ml}(s)$ and $D_{cl}(s)D_{ml}(s) + N_{cl}(s)N_{ml}(s)$ are Hurwitz, then it stabilizes P as well both in open- and closed-loop, i.e., $D_{cl}(s)D(s)$ and $D_{cl}(s)D(s) + N_{cl}(s)N(s)$ are Hurwitz.

For simplicity, denote

$$C_l P = \frac{N_{cl}(s)}{D_{cl}(s)} \frac{N(s)}{D(s)} = \frac{N_l(s)}{D_l(s)} \tag{7.39}$$

where $C_l P$ is the open-loop transfer function of Fig. 7.10. Correspondingly,

$$\frac{C_l P}{1 + C_l P} = \frac{N_l(s)}{N_l(s) + D_l(s)} \tag{7.40}$$

is the closed-loop transfer function of Fig. 7.10.

Based on these notations and VES III as shown in Fig. 7.9, we have the following result.

7.4.2 *Main Result*

Theorem 7.4 *The TSFC system as shown in Fig. 7.1(b) for an LTI plant is stable and tracking the reference signal, if only the following conditions are satisfied:*

(1) Each local controller stabilizes both its corresponding model and the true plant in the open-loop, i.e., $D_l(s), l \in L$ is Hurwitz;
(2) Each local controller stabilizes both its corresponding local model and the true plant in the closed-loop, i.e., $D_l(s) + N_l(s), l \in L$ is Hurwitz; and the resulting closed-loop control system is tracking the reference signal $y_r(t)$, i.e.,

$$\lim_{s\to 0}[D_l(s) + N_l(s)] = \lim_{s\to 0} N_l(s), l \in L$$

(3) $\forall\omega \in \mathbb{R}, 0 \le \mu_l \le 1, l \in L$

$$\begin{aligned}&\mu_1 D_2(j\omega)D_3(j\omega)\ldots[D_1(j\omega) + N_1(j\omega)]+\\&\mu_2 D_1(j\omega)D_3(j\omega)\ldots[D_2(j\omega) + N_2(j\omega)] + \cdots \neq 0\end{aligned} \tag{7.41}$$

Proof Consider the closed-loop transfer function of VES III (Fig. 7.9), we have

$$G(s) = \frac{Y(s)}{Y_r(s)} = \frac{\left(\sum_{l=1}^{m} \mu_l C_l\right) P}{1 + \left(\sum_{l=1}^{m} \mu_l C_l\right) P} \tag{7.42}$$

Taking (7.40) into account, we obtain

$$G(s) = \frac{\mu_1 N_1 D_2 D_3 \cdot + \mu_2 N_2 D_1 D_3 \cdot + \cdots}{\mu_1 D_2 D_3 \cdot (D_1 + N_1) + \mu_2 D_1 D_3 \cdot (D_2 + N_2) + \cdots} \tag{7.43}$$

From Condition (1) and the former part of Condition (2) we know that $D_l(s)$ and $N_l(s) + D_l(s), l \in L$ are all Hurwitz polynomials. Consequently, each term of the denominator of $G(s)$ is Hurwitz.

Further, according to Zero Exclusion Theorem [36], from condition (3) we know that the denominator polynomial of $G(s)$ is Hurwitz for all $0 \le \mu_l \le 1, l \in L$, i.e., the VES III as shown in Fig. 7.9 is stable.

Next we turn to to check the tracking performance of VES III as shown in Fig. 7.9.

From the latter part of condition (2), we know that

$$D_l(s) + N_l(s) \to N_l(s), l \in L \tag{7.44}$$

Substituting this into (7.43) gives

$$\begin{aligned}\lim_{s\to 0} G(s) &= \frac{\mu_1 N_1 D_2 D_3 + \mu_2 N_2 D_1 D_3 + \cdots}{\mu_1 N_1 D_2 D_3 \cdots + \mu_2 N_2 D_1 D_3 + \cdots}\\&= 1\end{aligned} \tag{7.45}$$

Further we have

$$\lim_{s\to 0} Y(s) = \lim_{s\to 0} G(s) Y_r(s) = \lim_{s\to 0} Y_r(s) \tag{7.46}$$

Then by the final value theorem of Laplace Transform, we get

$$\lim_{t\to\infty} [y(t) - y_r(t)] = \lim_{s\to 0} s[Y(s) - Y_r(s)] = 0 \tag{7.47}$$

Thus we conclude that VES III (Fig. 7.9) is stable and tracking, so is the original TSFC system (Fig. 7.1b).
That completes the proof.

Remark 7.1 Conditions (1) and (2) of Theorem 7.4 are related to the simultaneous stabilization problem [37–41], which is not a trivial problem. It is hard to test those required conditions because the plant is unknown. Hence we propose an alternative way to judge the simultaneous stabilization problem from the viewpoint of system output difference. Two propositions will be given in Appendix D.

7.5 Stability Analysis for Linear Parameter Jumping Plant and Nonlinear Time-Varying Plant

This section considers the linear parameter jumping plant and nonlinear time-varying plant that can be approximated by multiple linear models. Four cases will be considered in this section.

7.5.1 VES of TSFC Systems

The block diagrams of TSFC systems of time-varying plant both in state feedback form and in output feedback form are shown in Fig. 7.11a, b, respectively, where $P(t)$ is time-varying.

Next we construct VESs for different cases.

For case 1, we consider that $P(t)$ is a linear parameter jumping plant, and the time period between two consecutive parameter jumps is long enough. At each parameter jump, there is a model, say $M_i, i \in L$, matches the true model of the plant. Besides, the fuzzy rules and membership functions are designed appropriately such that the normalized membership function $\mu_i, i \in L$ identifies the correct local model at each parameter jump. In such situation, the VES for the TSFC system (as shown in Fig. 7.11a) can be constructed as shown in Fig. 7.12, where $u_i(t)$ is the output signal of C_i, and denote $\Delta u(t) = u(t) - u_i(t)$.

For case 2 and case 3, we consider that $P(t)$ is linear parameter jumping plant or nonlinear time-varying plant, but it can be approximated by multiple local linear models with approximation error $e(t) = x(t) - x_m(t) = o(\beta + \|x(t)\|)$ or $\|e(t)\| \leq M < \infty$, respectively. $x_m(t)$ is the state vector generated by $M_i, i \in L$. Then we get VES II as shown in Fig. 7.13 for cases 2 and case 3.

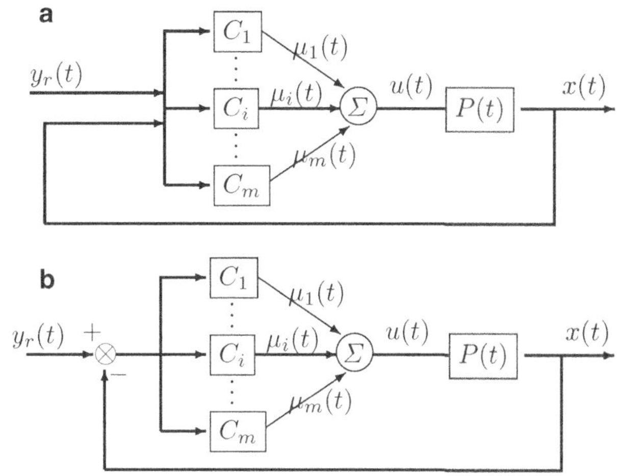

Fig. 7.11 **a** Block diagram of TSFC of time-varying plant: state feedback. **b** Block diagram of TSFC of time-varying plant: output feedback

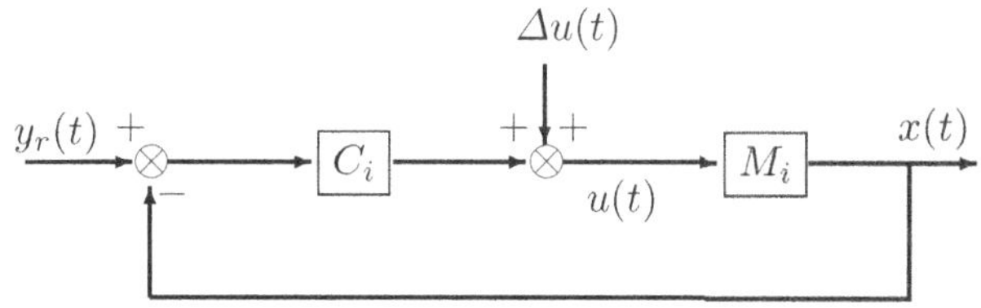

Fig. 7.12 VES I for TSFC of time-varying plant

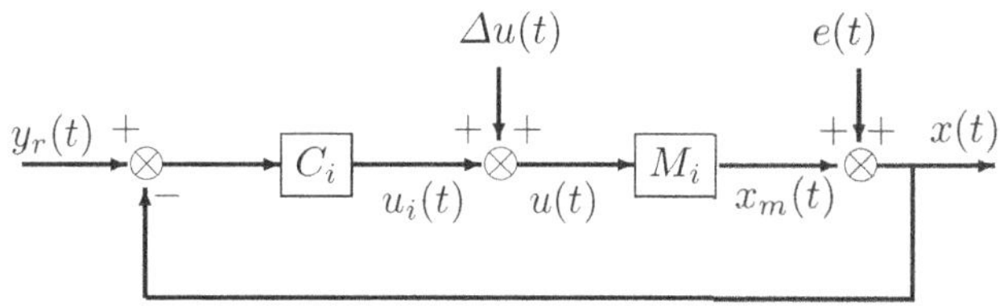

Fig. 7.13 VES II for TSFC of time-varying plant

For case 4, we consider that $P(t)$ is a linear parameter jumping plant, and the time period between two consecutive parameter jumps is long enough. At each parameter jump, there are some local models and corresponding local controllers, say M_i, $i \in L'$ and C_i, $i \in L'$, $L' := 1, 2, \ldots, m_i$, meet the requirements of Theorem 7.4. Then we get VES III for case 4 as shown in Fig. 7.14, where $\overline{C}_i = \sum_{j=1}^{m_i} \mu_j C_j$.

Next, we give the stability results based on VES I (Fig. 7.12), VES II (Fig. 7.13), and VES III (Fig. 7.14), respectively. To be specific, Theorem 7.5 will be proved based on VES I, Theorems 7.6 and 7.7 will be proved based on VES II, Theorem 5-4 will be proved based on VES III.

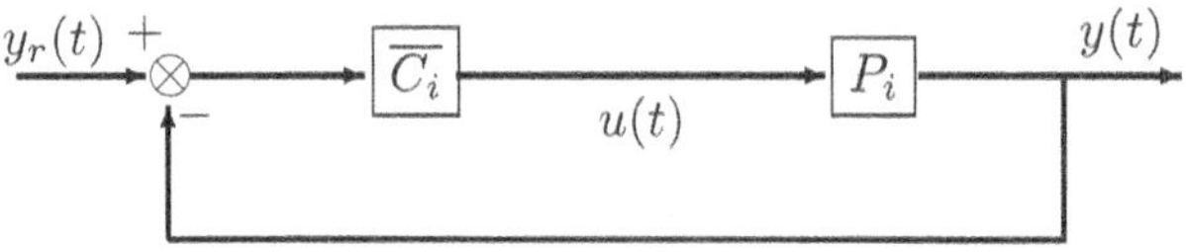

Fig. 7.14 VES III for TSFC of time-varying plant

7.5.2 Main Results

For case 1, we have the following result.

Theorem 7.5 *The TSFC system for a linear parameter jumping plant as shown in Fig. 7.11a is stable and tracking the reference signal, if only the following conditions are satisfied*

(1) Local model set in fuzzy rules covers the uncertainties of the plant, i.e., for each parameter jump, there is a model, say M_i, $i \in L$, matches the true model of the plant;
(2) Fuzzy rules and the membership functions correctly "identify" the true model of the plant at each parameter jump, i.e., if the subscripts of M_i and C_i are "frozen", the normalized membership function converge correctly in the sense that $\mu_i \to 1, \mu_j \to 0, j \neq i$;
(3) Each local controller together with its corresponding local model formulates a stable and tracking closed-loop control system;
(4) The time interval between consecutive parameter jump is long enough.

Proof In Fig. 7.12, suppose the subscripts of M_i and C_i are "frozen", then by conditions (1), (2), and (3), we know that the system as shown in Fig. 7.12 is equivalent, in the input-output sense, to the system as shown in Fig. 7.2. Thus, according to Theorem 7.1, it is clear that Fig. 7.12 with "frozen" subscript i represents a stable and tracking control system.
Further by condition (4), Fig. 7.12 with time-varying subscript i represents a "slow" switched system and each component system is stable. Then we know by switched system theory [42, 43] that the system as shown in Fig. 7.12 is stable and tracking, so is the TSFC system of time-varying plant as shown in Fig. 7.11a.
That completes the proof.

For case 2, we have the following result.

Theorem 7.6 *The TSFC system for a linear parameter jumping plant or a nonlinear time-varying plant, as shown in Fig. 7.11a, is stable and tracking the reference signal, if only the following conditions are satisfied:*

(1) Local model set in fuzzy rules covers the uncertainties of the plant, i.e., the plant can be approximated by multiple local models, say M_i, $i \in L$, with approximation error $e(t) = o(\beta + \|x(t)\|)$;

(2) *Fuzzy rules and the membership functions correctly "identify" the closest model(s) of the plant, i.e., if the subscripts of M_i and C_i are "frozen", the normalized membership function converge correctly in the sense that $\mu_i \to 1, \mu_j \to 0, j \neq i$;*
(3) *Each local controller together with its corresponding local model formulates a stable and tracking closed-loop control system;*
(4) *The time interval between consecutive parameter jump is long enough.*

Proof In Fig. 7.13, suppose subscripts of M_i and C_i are "frozen", then by conditions (1), (2), and (3), we know that the system as shown in Fig. 7.13 is equivalent, in the input-output sense, to the system as shown in Fig. 7.3 with $e(t) = o(\beta + \|x(t)\|)$. Thus, according to Theorem 7.2, Fig. 7.13 with "frozen" subscript i represents a stable and tracking closed-loop system.

Further by condition (4), Fig. 7.13 with time-varying subscript i is a "slow" switched system and each component system is stable. Then we know by switched system theory that the system as shown in Fig. 7.13 with $e(t) = o(\beta + \|x(t)\|)$ is stable and tracking, so is the TSFC system of time-varying plant (Fig. 7.11a).

That completes the proof.

For case 3, we have the following result.

Theorem 7.7 *The TSFC system for a linear parameter jumping plant or a nonlinear time-varying plant, as shown in Fig. 7.11a, is stable, if only the following conditions are satisfied:*

(1) *Local model set in fuzzy rules covers the uncertainties of the plant, i.e., the plant can be approximated by multiple local models, say $M_i, i \in L$, with approximation error $\|e(t)\| \leq M < \infty$;*
(2) *Fuzzy rules and the membership functions correctly "identify" the closest model(s) of the plant, i.e., if the subscripts of M_i and C_i are "frozen", the normalized membership function converge correctly in the sense that $\mu_i \to 1, \mu_j \to 0, j \neq i$;*
(3) *Each local controller together with its corresponding local model formulates a stable and tracking closed-loop control system;*
(4) *The time interval between consecutive parameter jump is long enough.*

Proof Consider the system as shown in Fig. 7.13, and suppose the subscripts of M_i and C_i are "frozen" then by conditions (1), (2), and (3), we know that the system is equivalent, in the input-output sense, to the system as shown in Fig. 7.3 with $\|e(t)\| \leq M < \infty$. Thus, according to Theorem 7.3, Fig. 7.13 with "frozen" subscript i is a stable closed-loop system.

Further by condition (4), Fig. 7.13 with time-varying subscript i represents a "slow" switched system and each component system is stable. Then we know by switched system theory that the system as shown in Fig. 7.13 with $\|e(t)\| \leq M < \infty$ is stable, so is the TSFC system of time-varying plant as shown in Fig. 7.11a.

That completes the proof.

For case 4, we have the following results.

Theorem 7.8 *The TSFC system for a linear parameter jumping plant, as shown in Fig. 7.11b, is stable and tracking the reference signal, if only the following conditions are satisfied:*

(1) For each parameter jump, there are some local models and corresponding controllers, say $M_i, i \in L'$ and $C_i, i \in L'$, satisfies the conditions of Theorem 7.4.
(2) The time interval between consecutive parameter jump is long enough.

Proof Considering Fig. 7.14, suppose the subscripts of P_i and $\overline{C}_i$ are "frozen", then by Condition (1), we know that the system as shown in Fig. 7.14 is equivalent to the system as shown in Fig. 7.9. Thus, according to Theorem 7.4, Fig. 7.14 with "frozen" subscript i represents a stable and tracking closed-loop system.

Further by condition (2), the system as shown in Fig. 7.14 with time-varying subscript i is a "slow" switched system and each component system is stable and tracking. Then we know by switched system theory that the system as shown in Fig. 7.14 is stable and tracking, so is the TSFC system of time-varying plant (Fig. 7.11b).

That completes the proof.

Remark 7.2 Although we only considered SISO TSFC systems, it is straight forward to extend the corresponding results to multi-input multi-output (MIMO) system except for Theorems 7.4 and 7.8, because we adopted state space model and norm operations to develop those theorems, i.e., Theorems 7.1, 7.2, 7.3, and 7.5, 7.6, 7.7.

7.6 Conclusion

With the help of VES concept and methodology, the stability problem of TSFC systems of continuous-time plants, including linear time-invariant plant, parameter jumping plant, and nonlinear time-varying plant, have been addressed in a unified framework. The main results obtained in this chapter demonstrate that the stability of a general TSFC system mainly depends on the following conditions:

(1) The local model set covers the uncertainties of the plant, i. e., the true model of plant is included in the model set, or there is a closest one in the model set with small enough modeling error;
(2) The fuzzy rules and the membership functions correctly 'identify' the model(s) of the plant;
(3) Each local controller together with its corresponding local model formulates a closed-loop stable control system and tracking the reference input signal.

These conditions are closely and reasonably related to each design procedure of a TSFC system, i.e., the accuracy of modeling and membership function. In other words, if only the fuzzy rules (including membership function) and models are designed/constructed properly, then a stable closed-loop TSFC system consisted of the parallel distributed compensator and the real plant, can be readily expected.

Although only TSFC systems are considered, we believe that VES approach could be used for stability analysis of any model based control systems, such as, model predictive control (MPC), active disturbance rejection control (ADRC) [48].

References

1. R.E. Precup, H. Hellendoorn, A survey on industrial applications of fuzzy control. Comput. Ind. **62**, 213–226 (2011)
2. T. Takagi, M. Sugeno, Fuzzy identification of systems and its applications to modeling and control. IEEE Trans. Syst. Man Cybern. **15**(1), 116–132 (1985)
3. M. Sugeno, On stability of fuzzy systems expressed by fuzzy rules with singleton consequents. IEEE Trans. Fuzzy Syst. **7**(2), 201–224 (1999)
4. A. Sala, T.M. Guerra, R. Babuska, Perspectives of fuzzy systems and control. Fuzzy Sets Syst. **156**, 432–444 (2005)
5. G. Feng, A survey on analysis and design of model-based fuzzy control systems. IEEE Trans. Fuzzy Syst. **14**(5), 676–697 (2006)
6. S.W. Cao, N.W. Rees, G. Feng, Analysis and design for a class of complex control systems-Part I: fuzzy modeling and identification. Automatica **33**, 1017–1028 (1997)
7. C. Fantuzzi, R. Rovatti, On the approximation capabilities of the homogeneous Takagi-Sugeno model, in *Proceedings of 5th IEEE International Conference on Fuzzy Systems* (New Orleans, LA, 1996), pp. 1067–1072
8. T.A. Johansen, R. Shorten, R. Murray-Smith, On the interpretation and identification of dynamic Takagi-Sugeno models. IEEE Trans. Fuzzy Syst. **8**(3), 297–313 (2000)
9. K. Tanaka, H.O. Wang, *Fuzzy Control Systems Design and Analysis: A LMI Approach* (Wiley, New York, 2001)
10. H. Ying, General SISO Takagi-Sugeno fuzzy systems with linear rule consequent are universal approximators. IEEE Trans. Fuzzy Syst. **6**(4), 582–587 (1998)
11. K. Zeng, N.Y. Zhang, W.L. Xu, A comparative study on sufficient conditions for Takagi-Sugeno fuzzy systems as universal approximators. IEEE Trans. Fuzzy Syst. **8**(6), 773–780 (2000)
12. K. Tanaka, M. Sugeno, Stability analysis and design of fuzzy control systems. Fuzzy Sets Syst. **12**, 135–156 (1992)
13. S.G. Cao, N.W. Rees, G. Feng, Stability analysis and design for a class of continuous-time fuzzy control systems. Int. J. Control **64**, 1069–1087 (1996)
14. S.G. Cao, N.W. Rees, G. Feng, H^{∞} control of nonlinear continuous-time systems based on dynamic fuzzy models. Int. J. Syst. Sci. **27**(9), 821–830 (1996)
15. S.G. Cao, N.W. Rees, G. Feng, Analysis and design for a class of complex control systems-Part II: Fuzzy controller design, Automatica **33**, 1029–1039 (1997)
16. D.J. Choi, P.G. Park, H-infinity state-feedback controller design for discrete-time fuzzy systems using fuzzy weighting-dependent Lyapunov functions. IEEE Trans. Fuzzy Syst. **11**(2), 271–278 (2003)
17. T.M. Guerra, L. Vermeiren, LMI-based relaxed nonquadratic stabilization conditions for non-linear systems in the Takagi-Sugeno's form. Automatica **40**(5), 823–829 (2004)
18. K. Tanaka, T. Hori, H.O. Wang, A multiple Lyapunov function approach to stabilization of fuzzy control systems. IEEE Trans. Fuzzy Syst. **11**(4), 582–589 (2003)

19. Y. Wang, Z.Q. Sun, F.C. Sun, Stability analysis and control of discrete-time fuzzy systems: a fuzzy Lyapunov function approach, in *Proceedings of 5th Asian Control Conference* (Melbourne, Australia, 2004), pp. 1855-1860
20. S.S. Zhou, G. Feng, J. Lam, S.Y. Xu, Robust H-infinity control for discrete fuzzy systems via basis-dependent Lyapunov functions. Inform. Sci. **174**(3–4), 197–217 (2005)
21. W. Zhang, On the stability and convergence of self-tuning control-virtual equivalent system approach. Int. J. Control **83**(5), 879–896 (2010)
22. W. Zhang, Stable weighted multiple model adaptive control: discrete-time stochastic plant. Int. J. Adapt. Control Signal Process. **27**(7), 562–581 (2013)
23. W. Zhang, Further results on stable weighted multiple model adaptive control: discrete-time stochastic plant. Int. J. Adapt. Control Signal Process. **29**(12), 1497–1514 (2015)
24. W. Zhang, X. Wang, Stability of T-S model based fuzzy control systems-virtual equivalent system approach, *International Conference on Computational Problem-Solving (ICCP)* (Sichuan, China, 2013)
25. Q. Li, W. Zhang, Y. Xu, S. Wang, Virtual equivalent system and stability criteria for discrete-time T-S model based fuzzy control systems, *24th Chinese Control and Decision Conference (CCDC)* (Taiyuan, China, 2012)
26. D.T. Magill, Optimal adaptive estimation of sampled stochastic processes. IEEE Trans. Autom. Control **10**, 434–439 (1965)
27. D.G Lainiotis, Partitioning: aunifying framework for adaptive systems I: Estimation; II: Control, in *Proceedings of IEEE*, vol. 64 (1976) , pp. 1126–1143 and 1182–1197
28. M. Athans et al., The stochastic control of the F-8C aircraft using a multiple model adaptive control (MMAC) method-Part I: Equilibrium flight. IEEE Trans. Autom. Control **22**, 768–780 (1977)
29. S. Fekri, M. Athans, A. Pascoal, Issues, progress and new results in robust adaptive control. Int. J. Adapt. Control Signal Process **20**(10), 519–579 (2006)
30. J. Shamma, M. Athans, Analysis of gain scheduled control for nonlinear plants. IEEE Trans. Autom. Control **35**(8), 898–907 (1990)
31. W. Rugh, Analytical framework for gain scheduling. IEEE Control Syst. Mag. **11**(1), 79–84 (1991)
32. R. Nichols et al., Gain scheduling for H1 controllers: a flight control example. IEEE Trans. Control Syst. Technol. **1**(2), 69–79 (1993)
33. D. Stilwell, W. Rugh, Interpolation of observer state feedback controllers for gain scheduling. IEEE Trans. Autom. Control **44**(6), 1225–1229 (1999)
34. F. Blanchini, The gain scheduling and the robust state feedback stabilization problems. IEEE Trans. Autom. Control **45**(11), 2061–2070 (2000)
35. C.A. Desoer, Vidyasagar M, *Feedback Systems: Input-output Properties* (Academic Press, New York, 1975)
36. B.R. Barmish, *New Tools for Robustness of Linear Systems* (Macmillan, New York, 1994)
37. V. Blondel, *Simultaneous Stabilization of Linear Systems*, vol. 191, Lecture Notes in Control and Information Sciences (Springer, London, 1994)
38. J. Doyle, B. Francis, A. Tannenbaum, *Feedback Control Theory* (Macmillan Publishing Company, New York, 1991)
39. M. Vidyasagar, *Control System Synthesis: A Factorization Approach* (The MIT Press, Massachusetts, 1985)
40. J. Ackermann, *Uncertainty and Control*, vol. 70, Lecture Notes in Control and Information Sciences (Springer, Berlin, 1985)
41. A. Alos, Stabilization of a class of plants with possible loss of outputs and actuator failures. IEEE Trans. Autom. Control **28**(2), 231–233 (1983)
42. D. Liberzon, A.S. Morse, Basic problems in stability and design of switched systems. IEEE Control Syst. Mag. **19**, 59–70 (1999)
43. R. Shorten, F. Wirth, O. Mason, K. Wulff, C. King, Stability criteria for switched and hybrid systems. SIAM Rev. **49**, 545–592 (2007)

44. E.M. Mellouli, I. Boumhidi, Direct adaptive fuzzy sliding mode controller without reaching phase for an uncertain three-tank-system. Int. J. Model. Identif. Control **25**(4), 335-342 (2016)
45. F. Tay, Z. Man, Z. Cao, S. Khoo, C.P. Tan, Sliding mode-like learning control for SISO complex systems with T-S fuzzy models. Int. J. Model. Identif. Control **16**(4), 317-326 (2012)
46. Y.-H. Chang, Y.-Y. Wang, M.-H. Hung, P.-C. Chen, Fuzzy-scheduling integral control of induction servo motor with actuator saturation via LMI approach. Int. J. Comput. Appl. Technol. **27**(2/3), 107–118 (2006)
47. H.-K. Lam, A review on stability analysis of continuous-time fuzzy-model-based control systems: From membership-function-independent to membership-function-dependent analysis. Eng. Appl. Artif. Intell. **68**, 390–408 (2018)
48. W. Wei, W. Zhang, D. Li, Y. Zhang, On the stability of linear active disturbance rejection control: Virtual equivalent system approach, in *Proceedings of the 2015 Chinese Intelligent Systems Conference* (2015), pp. 295-306
49. P. Li, W. Zhang, Towards a unified stability analysis of continuous-time T-S model based fuzzy control-virtual equivalent system approach. Int. J. Model. Identif. Control **31**(2), 113–123 (2019)

Appendix A
Lemma and Proofs for Chap. 2

Lemma A.1 *For the STC system described in Chap. 2, we have* $\|\tilde{\phi}(k)\| = O(\|\tilde{\phi}(k-i)\|)$*, where i is a limited integer.*

Proof Recall the definition of $\tilde{\phi}(k)$, i.e., Eq. (2.16)

$$\tilde{\phi}(k) = [y(k), \ldots, y(k-n), u(k-1), \ldots, u(k-d-m), y_r(k), \ldots, y_r(k-s_3)] \tag{A.1}$$

Then we have

$$\begin{aligned}\tilde{\phi}(k-1) =& [y(k-1), \ldots, y(k-n-1), u(k-2), \ldots, u(k-d-m-1), \\ & y_r(k-1), \ldots, y_r(k-s_3-1)]\end{aligned} \tag{A.2}$$

By components comparison between $\tilde{\phi}(k)$ and $\tilde{\phi}(k-1)$, We obtain

$$\begin{aligned}\|\tilde{\phi}(k)|^2 = & \|\tilde{\phi}(k-1)\|^2 - y^2(k-n-1) + y^2(k) \\ & - u^2(k-d-m-1) + u^2(k-1) \\ & - y_r^2(k-s_3-1) + y_r^2(k)\end{aligned} \tag{A.3}$$

That means

$$\|\tilde{\phi}(k)\|^2 \leq \|\tilde{\phi}(k-1)\|^2 + y^2(k) + u^2(k-1) + y_r^2(k) \tag{A.4}$$

Taking account of Eqs. (2.2) and (2.3) and $d \geq 1$, it follows that

$$y^2(k) \leq \|\tilde{\phi}(k-1)\|^2 + u^2(k-1) \tag{A.5}$$

W. Zhang and Q. Li, *Virtual Equivalent System Approach for Stability Analysis of Model-based Control Systems*, https://doi.org/10.1007/978-981-15-5538-1

Substituting this into (A.4), we get

$$\|\tilde{\phi}(k)\|^2 = O(\|\tilde{\phi}(k-1)\|^2) + O(u^2(k-1)) + M, \quad 0 \le M < \infty \tag{A.6}$$

Next, according to Eqs. (2.8), (2.10) and the fact that $\|\phi_c(k)\| < \infty$, we see that

$$u^2(k-1) = O(\|\phi_c(k-1)\|^2) = O(\|\tilde{\phi}(k-1)\|^2) \tag{A.7}$$

Then we obtain

$$\|\tilde{\phi}(k)\|^2 = O(\|\tilde{\phi}(k-1)\|^2) + M \tag{A.8}$$

Similarly, we have for limited integer i

$$\|\tilde{\phi}(k)\|^2 = O(\|\tilde{\phi}(k-i)\|^2) + M$$

Lemma A.2 *Given* $\tilde{\phi}(k) = \tilde{\phi}_1(k) + \tilde{\phi}_2(k)$, $\|\tilde{\phi}_1(k)\| < \infty$, $\|\tilde{\phi}_2(k)\| = o(\alpha + \|\tilde{\phi}(k)\|)$, *Then we have* $\|\tilde{\phi}(k)\| < \infty$

Proof Suppose, by contradiction, that $\|\tilde{\phi}(k)\|$ is unbounded. There must exists a subsequence $\|\tilde{\phi}(p_k)\|$ that goes to infinity. It is obvious that

$$\sum_{k=1}^{k=N} \|\tilde{\phi}(k)\| \le \sum_{k=1}^{k=N} \|\tilde{\phi}_1(k)\| + \sum_{k=1}^{k=N} \|\tilde{\phi}_2(k)\| \tag{A.9}$$

$$\sum_{k=1}^{k=N} \|\tilde{\phi}(p_k)\| \le \sum_{k=1}^{k=N} \|\tilde{\phi}_1(p_k)\| + \sum_{k=1}^{k=N} \|\tilde{\phi}_2(p_k)\| \tag{A.10}$$

Step 1, let's examine the first term in the right hand side of (A.10). Considering $\|\tilde{\phi}_1(k)\| < \infty$, we know that

$$\frac{\|\tilde{\phi}_1(p_k)\|}{\alpha + \|\tilde{\phi}(p_k)\|} \to 0$$

This together with Stolz Theorem leads to

$$\frac{\sum_{k=1}^{k=N} \|\tilde{\phi}_1(p_k)\|}{\sum_{k=1}^{k=N} (\alpha + \|\tilde{\phi}(p_k)\|)} \to 0 \tag{A.11}$$

Step 2, we estimate the second term in the right hand side of (A.10). Noticing the fact that $\|\tilde{\phi}(k)\| = 0, k \le 0$, then it follows that

$$\sum_{k=1}^{k=N} (\alpha + \|\tilde{\phi}(k-l)\|) \le \sum_{k=1}^{k=N} (\alpha + \|\tilde{\phi}(k)\|) \tag{A.12}$$

Then, by Stolz Theorem, we know the following inequality holds

$$0 \leq \frac{\sum_{k=1}^{k=N} \|\tilde{\phi}_2(k)\|}{\sum_{k=1}^{k=N} (\alpha + \|\tilde{\phi}(k)\|)} \leq \frac{\sum_{k=1}^{k=N} \|\tilde{\phi}_2(k)\|}{\sum_{k=1}^{k=N} (\alpha + \|\tilde{\phi}(k-l)\|)} \to 0 \tag{A.13}$$

According to Squeezing Principle, we get

$$\frac{\sum_{k=1}^{k=N} \|\tilde{\phi}_2(k)\|}{\sum_{k=1}^{k=N} (\alpha + \|\tilde{\phi}(k)\|)} \to 0 \tag{A.14}$$

Recalling the fact that the limits of sequence and its subsequences are the same, we have

$$\frac{\sum_{k=1}^{k=N} \|\tilde{\phi}_2(p_k)\|}{\sum_{k=1}^{k=N} (\alpha + \|\tilde{\phi}(p_k)\|)} \to 0 \tag{A.15}$$

Step 3, putting (A.11), (A.15) together with (A.10), and once again by Squeezing Principle, we get

$$\lim_{N\to\infty} \frac{\sum_{k=1}^{k=N} \|\tilde{\phi}(p_k)\|}{\sum_{k=1}^{k=N} (\alpha + \|\tilde{\phi}(p_k)\|)} = 0 \tag{A.16}$$

This contradicts the fact that

$$\lim_{N\to\infty} \frac{\sum_{k=1}^{k=N} \|\tilde{\phi}(p_k)\|}{\sum_{k=1}^{k=N} (\alpha + \|\tilde{\phi}(p_k)\|)} = 1 \tag{A.17}$$

Thus, we know the assumption, that $\|\tilde{\phi}(k)\|$ is unbounded, is false. That means

$$\|\tilde{\phi}(k)\| < \infty$$

Lemma A.3 *Virtual equivalent system, as shown in Fig. 2.10, can be decomposed into three subsystems, as shown in Figs. 2.11, 2.12, and 2.13, i.e., the input and output of the virtual equivalent system are identical to the sum of the inputs and outputs of the decomposed subsystems.*

Proof We give the proof by mathematical induction.

First, we know that for $k \leq 0$,

$$y(k) = y_1(k) + y_2(k) + y_3(k), \quad u(k) = u_1(k) + u_2(k) + u_3(k) \tag{A.18}$$

Second, assume (A.18) holds for $k, k-1, \ldots, 1$.

Considering Figs.2.10, 2.11, 2.12, and 2.13, respectively, we have

$$y(k+1) = \phi^T(k-d+1)\hat{\theta}_{t_k} + e_i(k+1) \tag{A.19}$$

$$y_1(k+1) = \phi_1^T(k-d+1)\hat{\theta}_{t_k} \tag{A.20}$$

$$y_2(k+1) = \phi_2^T(k-d+1)\hat{\theta}_{t_k} + e_i(k+1) \tag{A.21}$$

$$y_3(k+1) = \phi_3^T(k-d+1)\hat{\theta}_{t_k} \tag{A.22}$$

where $\hat{\theta}_{t_k}$ is the vector form of $P(t_k)$

$$\phi_1^T(k-d+1) = [y_1(k), \ldots, y_1(k-n+1), u(k-d+1), \ldots, u_1(k-d-m+1)] \tag{A.23}$$

$$\phi_2^T(k-d+1) = [y_2(k), \ldots, y_2(k-n+1), u_2(k-d+1), \ldots, u_2(k-d-m+1)] \tag{A.24}$$

$$\phi_3^T(k-d+1) = [y_3(k), \ldots, y_3(k-n+1), u_3(k-d+1), \ldots, u_3(k-d-m+1)] \tag{A.25}$$

According to the assumption, it is obvious that

$$\phi_1^T(k-d+1) + \phi_2^T(k-d+1) + \phi_3^T(k-d+1) = \phi^T(k-d+1) \tag{A.26}$$

Thus, we obtain

$$y_1(k+1) + y_2(k+1) + y_3(k+1) = \phi^T(k-d+1)\hat{\theta}_{t_k} + e_i(k+1) = y(k+1) \tag{A.27}$$

Next we observe $u_1(k+1)$, $u_2(k+1)$ and $u_3(k+1)$.

From Figs.2.10, 2.11, 2.12, and 2.13, respectively, we have

$$u(k+1) = \phi_c^T(k+1)\theta_c(t_{k+1}) + \Delta u'(k+1) \tag{A.28}$$

$$u_1(k+1) = \phi_{c1}^T(k+1)\theta_c(t_{k+1}) \tag{A.29}$$

$$u_2(k+1) = \phi_{c2}^T(k+1)\theta_c(t_{k+1}) \tag{A.30}$$

$$u_3(k+1) = \phi_{c3}^T(k+1)\theta_c(t_{k+1}) + \Delta u'(k+1) \tag{A.31}$$

where $\theta_c(t_{k+1})$ is the vector form of $C_{t_{k+1}}$.

$$\begin{aligned}\phi_c^T(k+1) =& [y(k+1), y(k), \ldots, y(k-s_1+1), u(k), u(k-1), \ldots,\\ & u(k-s_2+1), y_r(k+1), \ldots, y_r(k-s_3+1)]\end{aligned} \tag{A.32}$$

$$\begin{aligned}\phi_{c1}^T(k+1) =& [y_1(k+1), y_1(k), \ldots, y_1(k-s_1+1), u_1(k), u_1(k-1), \ldots,\\ & u_1(k-s_2+1), y_r(k+1), \ldots, y_r(k-s_3+1)]\end{aligned} \tag{A.33}$$

$$\begin{aligned}\phi_{c2}^T(k+1) =& [y_2(k+1), y_2(k), \ldots, y_2(k-s_1+1), u_2(k), u_2(k-1), \ldots,\\ & u_2(k-s_2+1), 0, \ldots, 0]\end{aligned} \tag{A.34}$$

$$\begin{aligned}\phi_{c3}^T(k+1) =& [y_3(k+1), y_3(k), \ldots, y_3(k-s_1+1), u_3(k), u_3(k-1), \ldots,\\ & u_3(k-s_2+1), 0, \ldots, 0]\end{aligned} \tag{A.35}$$

Based on (A.27), it is obvious that

$$\phi_{c1}^T(k+1) + \phi_{c2}^T(k+1) + \phi_{c3}^T(k+1) = \phi_c^T(k+1) \tag{A.36}$$

Then we get

$$u_1(k+1) + u_2(k+1) + u_3(k+1) = \phi_c^T(k+1)\theta_c(t_{k+1}) = u(k+1) \tag{A.37}$$

Consequently, we know that (A.18) holds for arbitrary k.

Lemma A.4 *Given*

$$\frac{1}{n}\sum_{k=1}^{n}|y_2(k)|^2 = o(1), \quad \lim_{n\to\infty} sup\frac{1}{n}\sum_{k=1}^{n}|y_1(k) - y_r(k)|^2 < \infty$$

Then we have

$$\frac{1}{n}\sum_{k=1}^{n}|y_1(k) - y_r(k) + y_2(k)|^2 \to \frac{1}{n}\sum_{k=1}^{n}|y_1(k) - y_r(k)|^2$$

Proof

$$\begin{aligned}&\frac{1}{n}\sum_{k=1}^{n}|y_1(k) - y_r(k) + y_2(k)|^2\\ &= \frac{1}{n}\sum_{k=1}^{n}[y_1(k) - y_r(k)]^2 + \frac{1}{n}\sum_{k=1}^{n}[y_2(k)]^2 + \frac{2}{n}\sum_{k=1}^{n}[y_1(k) - y_r(k)]y_2(k)\\ &\to \frac{1}{n}\sum_{k=1}^{n}|y_1(k) - y_r(k)|^2 + \frac{2}{n}\sum_{k=1}^{n}[y_1(k) - y_r(k)]y_2(k)\end{aligned} \tag{A.38}$$

By the Cauchy inequality, we know that

$$0 \le \left\{\frac{1}{n}\sum_{k=1}^{n}[y_1(k) - y_r(k)]y_2(k)\right\}^2 \le \left\{\frac{1}{n}\sum_{k=1}^{n}[y_1(k) - y_r(k)]^2\right\}\cdot\left\{\frac{1}{n}\sum_{k=1}^{n}[y_2(k)]^2\right\} \to 0$$

Then according to Squeezing Principle, we obtain

$$\left\{\frac{1}{n}\sum_{k=1}^{n}[y_1(k) - y_r(k)]y_2(k)\right\}^2 \to 0 \tag{A.39}$$

That is

$$\frac{1}{n}\sum_{k=1}^{n}[y_1(k) - y_r(k)]y_2(k) \to 0$$

Substituting this into (A.38), we have

$$\frac{1}{n}\sum_{k=1}^{n}|y_1(k) - y_r(k) + y_2(k)|^2 \to \frac{1}{n}\sum_{k=1}^{n}|y_1(k) - y_r(k)|^2$$

The proof is completed.

Appendix B
Lemma and Proofs for Chap. 5

Lemma B.1 *The set-point tracking problem of subsystem 1 of VES III is equivalent to the stabilizing problem of the following switched autonomous system.*

$$\begin{cases} x'(k+1) = A^c_{\sigma(k)} x'(k) + v(k), \\ y(k) = \overline{C}_{\sigma(k)} x'(k) + \overline{C}_{\sigma(k)} \alpha_i + \omega(k). \end{cases} \tag{B.1}$$

where $\sigma(k)$ is a piecewise constant function of time k, called a switching signal, $\sigma(k) = i \in \{1, \ldots, N\}$

System (B.1) is a stable switched system with "dwell time"

$$\tau_D = \max_{i \in \{1,\ldots,N\}} \left\{ \frac{-\ln h_i}{\ln \lambda_i} \right\}$$

where $0 < \lambda_i < 1$ and $h_i \geq 1$ are determined by $\|A^c_i\|^k \leq h_i \lambda_i^k, i \in \{1, \ldots, N\}$.

Proof The proof is divided into 2 steps.

Step 1: We prove that the stability and tracking of subsystem 1 of VES III is identical to the stabilization of a standard switched system.

Each local model of (5.39) can be written as an n-dimension ($n = \max\{na, nb + d\}$) state space formulation, for details see [1].

$$\begin{cases} x(k+1) = \overline{M}_i x(k) + \overline{N}_i u(k) + \overline{K}_i \omega(k), \\ y(k) = \overline{C}_i x(k) + \omega(k). \end{cases} \tag{B.2}$$

W. Zhang and Q. Li, *Virtual Equivalent System Approach for Stability Analysis of Model-based Control Systems*, https://doi.org/10.1007/978-981-15-5538-1

$$\overline{M}_i = \begin{bmatrix} -a_1 & 1 & 0 & \dots & 0 \\ -a_2 & 0 & 1 & \dots & 0 \\ \vdots & \vdots & \vdots & \ddots & \vdots \\ -a_{n-1} & 0 & 0 & 0 & 1 \\ -a_n & 0 & 0 & 0 & 0 \end{bmatrix} \tag{B.3}$$

$$\overline{N}_i = \begin{bmatrix} 0 \dots 0 \; b_0 \; b_2 \dots b_{n-1} \; b_{nb} \end{bmatrix}^T \tag{B.4}$$

$$\overline{K}_i = \begin{bmatrix} -a_1 \; -a_2 \dots -a_{n-1} \; -a_n \end{bmatrix}^T \tag{B.5}$$

$$\overline{C}_i = \begin{bmatrix} 1 \; 0 \dots 0 \end{bmatrix} \tag{B.6}$$

A general feedback control law can be expressed by

$$u(k) = K_x x(k) + K_v y_r(k) \tag{B.7}$$

Integrating (B.7) into (B.2), we have

$$\begin{cases} x(k+1) = (\overline{M}_i + \overline{N}_i K_x)x(k) + \overline{N}_i K_v y_r(k) + \overline{K}_i \omega(k), \\ y(k) = \overline{C}_i x(k) + \omega(k). \end{cases} \tag{B.8}$$

Set $A_i^c = \overline{M}_i + \overline{N}_i K_x$, $v(k) = \overline{K}_i \omega(k)$. It is obvious that $Ev(k) = 0$, $E\|v(k)\|^2 < \infty$

Then we have

$$\begin{cases} x(k+1) = A_i^c x(k) + \overline{N}_i K_v y_r(k) + v(k), \\ y(k) = \overline{C}_i x(k) + \omega(k). \end{cases} \tag{B.9}$$

Applying z transformation to (B.9), we get

$$y(k) = \overline{C}_i (zI - A_i^c)^{-1} \overline{N}_i K_v y_r(k) + \overline{C}_i (zI - A_i^c)^{-1} v(k) + \omega(k) \tag{B.10}$$

Considering Condition (3) of Theorem 5.6, we know that $Ey(k) = y_r(k)$, i.e.,

$$\lim_{z \to 1} \overline{C}_i (zI - A_i^c)^{-1} \overline{N}_i K_v = \overline{C}_i (I - A_i^c)^{-1} \overline{N}_i K_v = 1 \tag{B.11}$$

Next, introducing $x(k) = x'(k) + \alpha_i$, we obtain

$$\begin{cases} x'(k+1) = A_i^c x'(k) + (A_i^c - I)\alpha_i + \overline{N}_i K_\upsilon y_r(k) + \upsilon(k), \\ y(k) = \overline{C}_i x'(k) + \overline{C}_i \alpha_i + \omega(k). \end{cases} \tag{B.12}$$

Choosing α_i, such that

$$(A_i^c - I)\alpha_i + \overline{N}_i K_\upsilon y_r(k) = 0 \tag{B.13}$$

That means

$$\alpha_i = (I - A_i^c)^{-1} \overline{N}_i K_\upsilon y_r(k) \tag{B.14}$$

Integrating (B.13) into (B.12) we get

$$\begin{cases} x'(k+1) = A_i^c x'(k) + \upsilon(k), \\ y(k) = \overline{C}_i x'(k) + \overline{C}_i \alpha_i + \omega(k). \end{cases} \tag{B.15}$$

Thus, the stability and tracking problem of system (B.2) is transformed into a stabilizing problem of system (B.15). It is because the stability of (B.15) means $E\{x'(k)\} = 0$, i. e., $E\{x(k)\} = \alpha_i$. Further, considering (B.11) and (B.14), we have $E\{y_r(k) - y(k)\} = y_r(k) - \overline{C}_i E\{x'(k)\} - \overline{C}_i \alpha_i = y_r(k) - \overline{C}_i (I - A_i^c)^{-1} \overline{N}_i K_\upsilon y_r(k) = 0$.

Consequently, the stability and tracking problem of subsystem 1 is identical to the stability of the following switched system.

$$\begin{cases} x'(k+1) = A_{\sigma(k)}^c x'(k) + \upsilon(k), \\ y(k) = \overline{C}_{\sigma(k)} x'(k) + \overline{C}_{\sigma(k)} \alpha_i + \omega(k). \end{cases} \tag{B.16}$$

where $\sigma(k) = i \in \{1, \ldots, N\}$

Step2: We give the "dwell time" of system described by (B.16). Along the same line of "dwell time" analysis for discrete-time switched systems in [2], we have

$$\tau_D = \max_{i \in \{1, \ldots, N\}} \{\frac{-\ln h_i}{\ln \lambda_i}\} \tag{B.17}$$

where $0 < \lambda_i < 1$ and $h_i \geq 1$ are determined by $\|A_i^c\|^k \leq h_i \lambda_i^k, i \in \{1, \ldots, N\}$.

That completes the proof.

Lemma B.2 *The sequence* $[\beta_i(k)]^{ceil(\frac{1}{1-\beta_i(k)})}$ *in Weighting algorithm 2 is monotonically increasing with upper bound* $1/e$.

Proof Recalling Weighting algorithm 2, it is obvious that

$$[\beta_i(k)]^{ceil\left(\frac{1}{1-\beta_i(k)}\right)} \leq [\beta_i(k)]^{\frac{1}{1-\beta_i(k)}} \tag{B.18}$$

Further we know that sequence $[\beta_i(k)]^{\frac{1}{1-\beta_i(k)}}$, $\beta_i(k) \uparrow 1$ can be rewritten as

$$\left[1 - \frac{1}{\alpha_i(k)}\right]^{\alpha_i(k)}, \alpha_i(k) \to \infty$$

which is monotonically increasing as $\alpha_i(k) \to \infty$.

Then we have

$$\left[1 - \frac{1}{\alpha_i(k)}\right]^{\alpha_i(k)} \leq \lim_{\alpha_i(k)\to\infty} [1 - \frac{1}{\alpha_i(k)}]^{\alpha_i(k)} = \frac{1}{e}$$

Thus we obtain

$$[\beta_i(k)]^{ceil\left(\frac{1}{1-\beta_i(k)}\right)} \leq [\beta_i(k)]^{\frac{1}{(1-\beta_i(k)}} \leq \frac{1}{e} \tag{B.19}$$

That completes the proof.

References

1. K.J. Åström, B. Wittenmark, *Adaptive Control* (Addison-Wesley, Reading, MA, 1995)
2. G. Zhai, B. Hu, K. Yasuda, A.N. Michel, Qualitative analysis of discrete-time switched systems, in *Proceedings of the American Control Conference*, vol. 3 (Anchorage, AK, 2002), pp. 1880–1885

Appendix C
Lemma and Proofs for Chap. 6

Lemma C.1 *For stable dynamic output feedback controllers $C_i, i = 1, 2, \ldots, m$, we have*

$$||x_{cj}(t)|| = O(||x_{ci}(t)||), i, j = 1, 2, \ldots, m, i \neq j \tag{C.1}$$

Proof Solving the state-space equation, we obtain

$$x_{ci}(t) = \int_0^t e^{A_{ci}(t-\tau)} B_{ci} y(\tau) d\tau \tag{C.2}$$

Since each controller C_i is stable, then there must exist constants $\lambda_i > 0, k_i > 0$, such that

$$\|e^{A_{ci}(t-\tau)}\| \leq k_i e^{-\lambda_i(t-\tau)} \tag{C.3}$$

Further by (C.2) we have

$$\begin{aligned} \|x_{ci}(t)\| &\leq k_i \|B_{ci}\| \int_0^t e^{-\lambda_i(t-\tau)} \|y(\tau)\| d\tau \\ &= \bar{k}_i \int_0^t e^{-\lambda_i(t-\tau)} \|y(\tau)\| d\tau \end{aligned} \tag{C.4}$$

where $\bar{k}_i = k_i \|B_{ci}\|$.

W. Zhang and Q. Li, *Virtual Equivalent System Approach for Stability Analysis of Model-based Control Systems*, https://doi.org/10.1007/978-981-15-5538-1

Similarly we have

$$\begin{aligned}\|x_{cj}(t)\| &\leq \bar{k}_j \int_0^t e^{-\lambda_j(t-\tau)}\|y(\tau)\|d\tau \\ &= \bar{k}_i \frac{\bar{k}_j}{\bar{k}_i} \int_0^t e^{-\lambda_i \frac{\lambda_j}{\lambda_i}(t-\tau)}\|y(\tau)\|d\tau\end{aligned} \tag{C.5}$$

Let $\frac{\lambda_j}{\lambda_i}(t-\tau) = t-\tau'$, we get

$$\begin{aligned}\|x_{cj}(t)\| &\leq \bar{k}_i \frac{\bar{k}_j}{\bar{k}_i}\frac{\lambda_i}{\lambda_j} \int_{(\frac{\lambda_j}{\lambda_i}-1)t}^t e^{-\lambda_i(t-\tau')}\|y(\tau')\|d\tau' \\ &= \bar{k}_i \frac{\bar{k}_j}{\bar{k}_i}\frac{\lambda_i}{\lambda_j} \int_{(\frac{\lambda_j}{\lambda_i}-1)t}^t e^{-\lambda_i(t-\tau)}\|y(\tau)\|d\tau\end{aligned} \tag{C.6}$$

Noticing the following facts

(1) $\frac{\bar{k}_j}{\bar{k}_i}\frac{\lambda_i}{\lambda_j}$ is a constant;

(2) $\int_{(\frac{\lambda_j}{\lambda_i}-1)t}^t e^{-\lambda_i(t-\tau)}\|y(\tau)\|d\tau = \int_0^t e^{-\lambda_i(t-\tau)}\|y(\tau)\|d\tau$, when $\frac{\lambda_j}{\lambda_i} < 1$;

(It is because $y(\tau) = 0$, for $\tau \leq 0$)

(3) $\int_{(\frac{\lambda_j}{\lambda_i}-1)t}^t e^{-\lambda_i(t-\tau)}\|y(\tau)\|d\tau \leq \int_0^t e^{\lambda_i(t-\tau)}\|y(\tau)\|d\tau$, when $\frac{\lambda_j}{\lambda_i} \geq 1$;

(It is because $e^{-\lambda_i(t-\tau)}\|y(\tau)\| \geq 0$, for $\tau \geq 0$)

then we have

$$\begin{aligned}\|x_{cj}(t)\| &\leq \bar{k}_i \frac{\bar{k}_j}{\bar{k}_i}\frac{\lambda_i}{\lambda_j} \int_0^t e^{-\lambda_i(t-\tau)}\|y(\tau)\|d\tau \\ &= \frac{\bar{k}_j}{\bar{k}_i}\frac{\lambda_i}{\lambda_j}\|x_{ci}(t)\|\end{aligned} \tag{C.7}$$

i. e.,

$$||x_{cj}(t)|| = O(||x_{ci}(t)||), i, j = 1, 2, \ldots, m, i \neq j$$

That completes the proof.

Lemma C.2 *Consider the following stable continuous-time LTI system*

$$\begin{cases} \dot{x}(t) = Ax(t) + Bu(t) \\ y(t) = Dx(t) + Eu(t) \\ x(0) = 0 \end{cases} \tag{C.8}$$

if $\lim_{t\to\infty} \|u(t)\| = 0$*, then*

$$\lim_{t\to\infty} \|x(t)\| = 0$$

$$\lim_{t\to\infty} \|y(t)\| = 0$$

Proof Considering $x(0) = 0$, solve the state equation of (C.8), we obtain

$$x(t) = \int_0^t e^{A(t-\tau)} Bu(\tau) d\tau \tag{C.9}$$

Since (C.8) represents a stable system, i. e., A is a Hurwitz matrix, there must exist constants $\lambda > 0$, $c > 0$, such that

$$\|e^{A(t-\tau)}\| \le ce^{-\lambda(t-\tau)} \tag{C.10}$$

Further we know that

$$0 \le \|x(t)\| \le c\|B\| \int_0^t e^{-\lambda(t-\tau)} \|u(\tau)\| d\tau \tag{C.11}$$

Considering Lemma C.4, we have

$$\lim_{t\to\infty} \int_0^t e^{-\lambda(t-\tau)} \|u(\tau)\| d\tau = 0 \tag{C.12}$$

Then according to Squeezing Principle, (C.11) and (C.12) yield

$$\lim_{t\to\infty} \|x(t)\| = 0$$

Further, from $y(t) = Dx(t) + Eu(t)$, we get

$$\lim_{t\to\infty} \|y(t)\| = 0$$

That completes the proof.

Corollary 1 *Consider a stable continuous-time LTI system as described in (C.8), if* $\|u(t)\| = o(\|f(t)\| + M)$*, then* $\|x(t)\| = o(\|f(t)\| + M)$*,* $\|y(t)\| = o(\|f(t)\| + M)$*, where* $f(t)$ *is a vector function,* $0 < M < \infty$*.*

Proof Dividing both sides of (C.8) by $(\|f(t)\| + M)$, then along the same lines of the proof of Lemma C.2, we get

$$\frac{\|x(t)\|}{\|f(t)\| + M} \to 0$$

$$\frac{\|y(t)\|}{\|f(t)\| + M} \to 0$$

i. e.,

$$\|x(t)\| = o(\|f(t)\| + M)$$

$$\|y(t)\| = o(\|f(t)\| + M)$$

That completes the proof.

Lemma C.3 *Given vectors* $x(t), x_1(t), x_2(t)$*, if*

$$x(t) = x_1(t) + x_2(t),\ \|x_1(t)\| < \infty, x_2(t) = o(\|x(t)\|)$$

then we have

$$\|x(t)\| < \infty$$

Proof It is obvious that

$$\begin{aligned} 0 \leq \|x(t)\| &\leq \|x_1(t)\| + \|x_2(t)\| \\ &= \|x_1(t)\| + o(\|x(t)\|) \end{aligned} \tag{C.13}$$

Next we prove the conclusion of the lemma by reduction to absurdity.

Suppose $\|x(t)\|$ is unbounded, there must exist a subseries $\|x(t_k)\|$ that goes to infinity, i. e, $\|x(t_k)\| \to \infty$, satisfying

$$0 \leq \|x(t_k)\| \leq \|x_1(t_k)\| + o(\|x(t_k)\|) \tag{C.14}$$

Dividing both sides of (C.14) by $\|x(t_k)\|$, we have

$$0 \leq \frac{\|x(t_k)\|}{\|x(t_k)\|} \leq \frac{\|x_1(t_k)\|}{\|x(t_k)\|} + o(1) \tag{C.15}$$

Considering $\|x_1(t_k)\|$ is bounded, then by Squeezing Principle, (C.15) gives a contradiction that

$$\frac{\|x(t_k)\|}{\|x(t_k)\|} \to 0 \tag{C.16}$$

Thus the assumption that $\|x(t)\|$ is unbounded can not hold. Then we have

$$\|x(t)\| < \infty \tag{C.17}$$

That completes the proof.

Lemma C.4 *Suppose* $\lambda > 0$ *is a constant,* $\lim_{t\to\infty} \|u(t)\| = 0$ *, then*

$$f(t) = \int_0^t e^{-\lambda(t-\tau)} \|u(\tau)\| d\tau \to 0, \textit{as } t \to \infty$$

Proof Let $f_1(t) = e^{-\lambda t}$, $f_2(t) = \|u(t)\| \to 0$, then $f(t)$ is the convolution of $f_1(t)$ and $f_2(t)$, thus, we have their Laplace transforms

$$\begin{aligned}
F_1(s) &= \mathcal{L}\{f_1(t)\} = \frac{1}{s+\lambda} \\
F_2(s) &= \mathcal{L}\{f_2(t)\} \\
F(s) &= \mathcal{L}\{f(t)\} = F_1(s) \cdot F_2(s)
\end{aligned}$$

By Final Value Theorem we have

$$\lim_{t\to\infty} f_2(t) = \lim_{s\to 0} sF_2(s) = 0$$

Further we have

$$\begin{aligned}
\lim_{t\to\infty} f(t) &= \lim_{s\to 0} sF_1(s) \cdot F_2(s) \\
&= \lim_{s\to 0} sF_2(s) \cdot \lim_{s\to 0} F_1(s) \\
&= 0 \cdot \left(\frac{1}{\lambda}\right) \\
&= 0
\end{aligned}$$

That completes the proof.

Appendix D
Lemma and Proofs for Chap. 7

D.1 Big-Oh and Little Oh

Definition of the Big-Oh:

We write $f = O(g)$, if there exists a constant c such that $|f(x)| \leq c|g(x)|$ for all $x \in X$, where X denotes a set of real numbers contained in the domain of the functions $f(x)$ and $g(x)$.

Definition of the Little-Oh:

Let $f, g : [0, x_0) \to \mathbb{R}$, g is positive-valued. We write $f = o(g)$ as $x \to x_1$, if and only if

$$\lim_{x \to x_1} \frac{|f(x)|}{g(x)} \to 0$$

where x_1 may be a finite real number or ∞.

D.2 Propositions for Simultaneous Stabilization Problem

Two propositions are given to address the on-line test of simultaneous stabilization of T-S local models and the real plant, according to local models output differences withe real plant, i.e., the differences between the calculated outputs of T-S local models and the measured output of the real plant.

Proposition 1 *Model M_i and plant P can be simultaneously stabilized by controller C_i, if their output difference in the closed-loop system (as shown in Fig. D.1) is $e(t) = y(t) - y_m(t) \to 0$ or $e(t) = o(\lambda + \|y(t)\|)$*

W. Zhang and Q. Li, *Virtual Equivalent System Approach for Stability Analysis of Model-based Control Systems*, https://doi.org/10.1007/978-981-15-5538-1

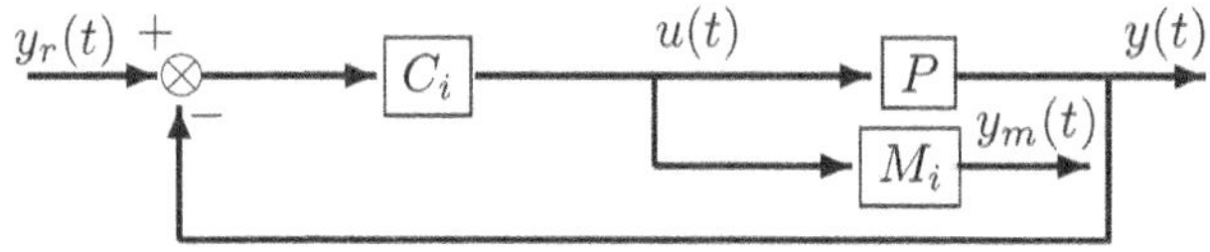

Fig. D.1 Block diagram for Proposition 1

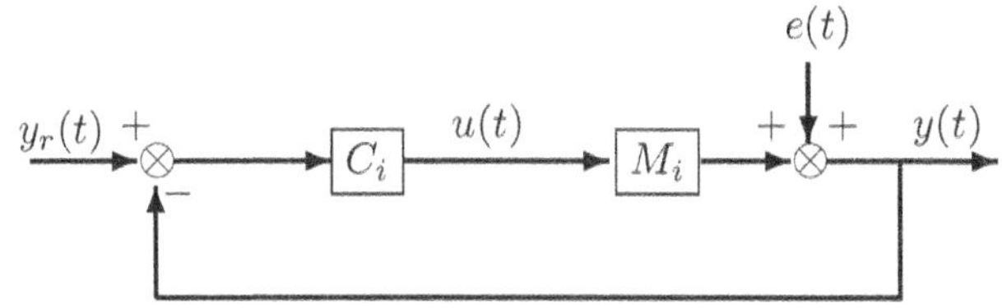

Fig. D.2 VES of Fig. D.1

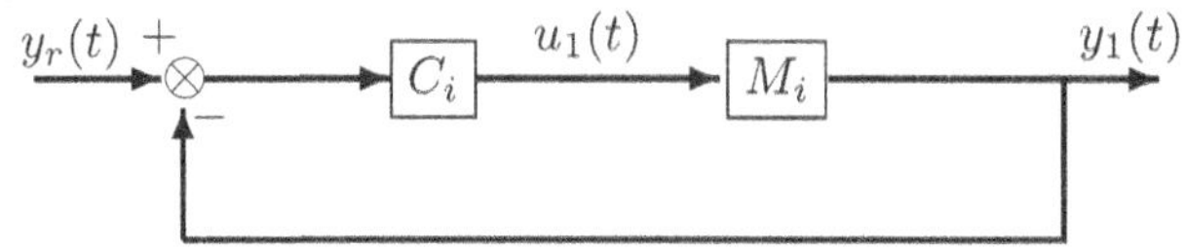

Fig. D.3 Decomposed subsystem 1 of Fig. D.2

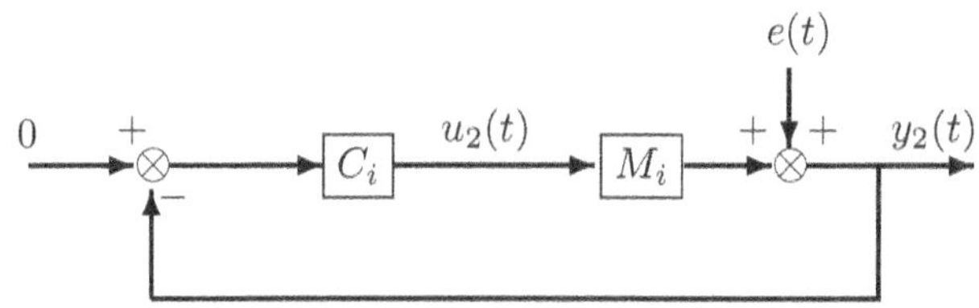

Fig. D.4 Decomposed subsystem 2 of Fig. D.2

Proof The block diagram of the related closed-loop control system is shown in Fig. D.1, and its VES can be constructed as shown in Fig. D.2.

It is easy to see that if C_i stabilizes M_i, then the VES as shown in Fig. D.2 can be decomposed into two stable subsystems, as shown in Figs. D.3, and D.4 with $e(t) \to 0$ or $e(t) = o(\lambda + \|y(t)\|)$, respectively.

Then following the same lines as in the proof of Theorems 7.1, 7.2, and 7.3, i.e., by reduction to absurdity, we obtain the conclusion that the VES as shown in Fig. D.2 is stable, so is the system as shown in Fig. D.1, i.e., C_i stabilizes P.

Similarly (thus the proof is omitted here), we have Proposition 2.

Proposition 2 *Model M_i and plant P can be simultaneously stabilized by controller C_i, if their output difference in the closed-loop system (as shown in Fig. D.1) is bounded, i.e.*

$$\|e(t)\| \leq M < \infty$$

The manufacturer's authorised representative in the EU is Springer Nature Customer Service Centre GmbH, Europaplatz 3, 69115 Heidelberg, Germany. If you have any concerns regarding our products, please contact ProductSafety@springernature.com

Printed and bound by CPI Group (UK) Ltd, Croydon, CR0 4YY
30/07/2026
02181693-0020